Lecture Notes in Computer Science 11514

Commenced Publication in 1973
Founding and Former Series Editors:
Gerhard Goos, Juris Hartmanis, and Jan van Leeuwen

More information about this series at http://www.springer.com/series/7409

Keke Chen · Sangeetha Seshadri ·
Liang-Jie Zhang (Eds.)

Big Data –
BigData 2019

8th International Congress
Held as Part of the Services Conference Federation, SCF 2019
San Diego, CA, USA, June 25–30, 2019
Proceedings

 Springer

Editors
Keke Chen ⓘ
Wright State University
Dayton, OH, USA

Sangeetha Seshadri
IBM Almaden Research Center
San Jose, CA, USA

Liang-Jie Zhang ⓘ
Kingdee
International Software Group Co., Ltd.
Shenzhen, China

ISSN 0302-9743 ISSN 1611-3349 (electronic)
Lecture Notes in Computer Science
ISBN 978-3-030-23550-5 ISBN 978-3-030-23551-2 (eBook)
https://doi.org/10.1007/978-3-030-23551-2

LNCS Sublibrary: SL3 – Information Systems and Applications, incl. Internet/Web, and HCI

This Springer imprint is published by the registered company Springer Nature Switzerland AG
The registered company address is: Gewerbestrasse 11, 6330 Cham, Switzerland

Preface

The 2019 International Congress on Big Data (BigData 2019) aimed to provide an international forum that formally explores various business insights of all kinds of value-added "services." Big data is a key enabler of exploring business insights and economics of services.

BigData 2019 was part of the Services Conference Federation (SCF). SCF 2019 had the following ten collocated service-oriented sister conferences: 2019 International Conference on Web Services (ICWS 2019), 2019 International Conference on Cloud Computing (CLOUD 2019), 2019 International Conference on Services Computing (SCC 2019), 2019 International Congress on Big Data (BigData 2019), 2019 International Conference on AI & Mobile Services (AIMS 2019), 2019 World Congress on Services (SERVICES 2019), 2019 International Congress on Internet of Things (ICIOT 2019), 2019 International Conference on Cognitive Computing (ICCC 2019), 2019 International Conference on Edge Computing (EDGE 2019), and 2019 International Conference on Blockchain (ICBC 2019). As the founding member of SCF, the First International Conference on Web Services (ICWS) was held in June 2003 in Las Vegas, USA. The First International Conference on Web Services—Europe 2003 (ICWS-Europe 2003) was held in Germany in Oct, 2003. ICWS-Europe 2003 is an extended event of the 2003 International Conference on Web Services (ICWS 2003) in Europe. In 2004, ICWS-Europe was changed to the European Conference on Web Services (ECOWS), which was held in Erfurt, Germany. To celebrate its 16th birthday, SCF 2018 was held successfully in Seattle, USA.

This volume presents the accepted papers for BigData 2019, held in San Diego, USA, during June 25–30, 2019. The major topics of BigData 2019 included but were not limited to: big data architecture, big data modeling, big data as a service, big data for vertical industries (government, healthcare, etc.), big data analytics, big data toolkits, big data open platforms, economic analysis, big data for enterprise transformation, big data in business performance management, big data for business model innovations and analytics, big data in enterprise management models and practices, big data in government management models and practices, and big data in smart planet solutions.

We accepted nine full papers. Each was reviewed and selected by at least three independent members of the BigData 2019 international Program Committee. We are pleased to thank the authors, whose submissions and participation made this conference possible. We also want to express our thanks to the Organizing Committee and Program Committee members, for their dedication in helping to organize the conference and in reviewing the submissions. We would like to thank Prof. Seong-Jong Park, who provided continuous support for this conference. We look forward to your great contributions as a volunteer, author, and conference participant for the fast-growing worldwide services innovations community.

May 2019   Keke Chen
                                                              Sangeetha Seshadri
                                                                  Liang-Jie Zhang

Organization

General Chair

Seong-Jong Park   Louisiana State University, USA

Program Chairs

Sangeetha Seshadri   IBM Almaden Research Center, USA
Keke Chen   Wright State University, USA

Services Conference Federation (SCF 2019)

SCF 2019 General Chairs

Calton Pu   Georgia Tech, USA
Wu Chou   Essenlix Corporation, USA
Ali Arsanjani   8x8 Cloud Communications, USA

SCF 2019 Program Chair

Liang-Jie Zhang   Kingdee International Software Group Co., Ltd., China

SCF 2019 Finance Chair

Min Luo   Services Society, USA

SCF 2019 Industry Exhibit and International Affairs Chair

Zhixiong Chen   Mercy College, USA

SCF 2019 Operations Committee

Huan Chen   Kingdee International Software Group Co., Ltd., China
Jing Zeng   Kingdee International Software Group Co., Ltd., China
Liping Deng   Kingdee International Software Group Co., Ltd., China
Yishuang Ning   Tsinghua University, China
Sheng He   Tsinghua University, China

SCF 2019 Steering Committee

Calton Pu (Co-chair)   Georgia Tech, USA
Liang-Jie Zhang (Co-chair)   Kingdee International Software Group Co., Ltd., China

Bigdata 2019 Program Committee

Hussein Al-Olimat	Wright State University, USA
Yan Bai	University of Washington Tacoma, USA
Shreyansh Bhatt	Wright State University, USA
Marios Dikaiakos	University of Cyprus, Cyprus
Wei Dong	Ann Arbor Algorithms Inc., USA
Manas Gaur	Wright State University, USA
Anastasios Gounaris	Aristotle University of Thessaloniki, Greece
Sheng He	Tsinghua University, China
Shihao Ji	Georgia State University, USA
Kosaku Kimura	Fujitsu Laboratories Ltd., Japan
Harald Kornmayer	DHBW Mannheim, Germany
Ugur Kursuncu	Wright State University, USA
Sarasi Lalithsena	IBM Silicon Valley, USA
Yu Liang	University of Tennessee at Chattanooga, USA
Xiaogang Ma	University of Idaho, USA
Jianyun Nie	University of Montreal, Canada
Hemant Purohit	George Mason University, USA
Sagar Sharma	Wright State University, USA
Luiz Angelo Steffenel	Université de Reims Champagne-Ardenne, France
Pierre Sutra	Telecom SudParis, France
Jianwu Wang	University of Maryland Baltimore County, USA
Wenbo Wang	GoDaddy, USA
Haruo Yokota	Tokyo Institute of Technology, Japan
Daqing Yun	Harrisburg University, USA

Contents

Designing and Implementing Data Warehouse for Agricultural Big Data

Vuong M. Ngo[✉], Nhien-An Le-Khac, and M-Tahar Kechadi

School of Computer Science, University College Dublin, Dublin, Ireland
{vuong.ngo,an.lekhac,tahar.kechadi}@ucd.ie

Abstract. In recent years, precision agriculture that uses modern information and communication technologies is becoming very popular. Raw and semi-processed agricultural data are usually collected through various sources, such as: Internet of Thing (IoT), sensors, satellites, weather stations, robots, farm equipment, farmers and agribusinesses, etc. Besides, agricultural datasets are very large, complex, unstructured, heterogeneous, non-standardized, and inconsistent. Hence, the agricultural data mining is considered as Big Data application in terms of volume, variety, velocity and veracity. It is a key foundation to establishing a crop intelligence platform, which will enable resource efficient agronomy decision making and recommendations. In this paper, we designed and implemented a continental level agricultural data warehouse by combining Hive, MongoDB and Cassandra. Our data warehouse capabilities: (1) flexible schema; (2) data integration from real agricultural multi datasets; (3) data science and business intelligent support; (4) high performance; (5) high storage; (6) security; (7) governance and monitoring; (8) consistency, availability and partition tolerant; (9) distributed and cloud deployment. We also evaluate the performance of our data warehouse.

Keywords: Data warehouse · Big Data · Precision agriculture

1 Introduction

In 2017 and 2018, annual world cereal productions were 2,608 million tons [30] and 2,595 million tons [7], respectively. However, there were also around 124 million people in 51 countries faced food crisis and food insecurity [8]. According to United Nations [29], we need an increase 60% of cereal production to meet 9.8 billion people needs by 2050. To satisfy the massively increase demand for food, crop yields must be significantly increased by using new farming approaches, such as precision agriculture. As reported in [6], precision agriculture is vitally important for the future and can make a significant contribution to food security and safety.

The precision agriculture's current mission is to use the decision-support system based on Big Data approaches to provide precise information for more control of farming efficiency and waste, such as awareness, understanding, advice, early warning, forecasting and financial services. An efficient agricultural data warehouse (DW) is required to extract useful knowledge and support decision-making. However, currently there are very few reports in the literature that

© Springer Nature Switzerland AG 2019
K. Chen et al. (Eds.): BigData 2019, LNCS 11514, pp. 1–17, 2019.
https://doi.org/10.1007/978-3-030-23551-2_1

focus on the design of efficient DWs with the view to enable Agricultural Big Data analysis and mining. The design of large scale agricultural DWs is very challenging. Moreover, the precision agriculture system can be used by different kinds of users at the same time, for instance by both farmers and agronomists. Every type of user needs to analyse different information sets thus requiring specific analytics. The agricultural data has all the features of Big Data:

1. Volume: The amount of agricultural data is rapidly increasing and is intensively produced by endogenous and exogenous sources. The endogenous data is collected from operation systems, experimental results, sensors, weather stations, satellites and farm equipment. The systems and devices in the agricultural ecosystem can connect through IoT. The exogenous data concerns the external sources, such as farmers, government agencies, retail agronomists and seed companies. They can help with information about local pest and disease outbreak tracking, crop monitoring, market accessing, food security, products, prices and knowledge.
2. Variety: Agricultural data has many different forms and formats, such as structured and unstructured data, video, imagery, chart, metrics, geo-spatial, multi-media, model, equation and text.
3. Velocity: The produced and collected data increases at high rate, as sensing and mobile devices are becoming more efficient and cheaper. The datasets must be cleaned, aggregated and harmonised in real-time.
4. Veracity: The tendency of agronomic data is uncertain, inconsistent, ambiguous and error prone because the data is gathered from heterogeneous sources, sensors and manual processes.

In this research, firstly, we analyze popular DWs to handle agricultural Big Data. Secondly, an agricultural DW is designed and implemented by combining Hive, MongoDB, Cassandra, and constellation schema on real agricultural datasets. Our DW has enough main features of a DW for agricultural Big Data. These are: (1) high storage, high performance and cloud computing adapt for the volume and velocity features; (2) flexible schema and integrated storage structure to adapt the variety feature; (3) data ingestion, monitoring and security adapt for the veracity feature. Thirdly, the effective business intelligent support is illustrated by executing complex HQL/SQL queries to answer difficult data analysis requests. Besides, an experimental evaluation is conducted to present good performance of our DW storage. The rest of this paper is organised as follows: in the next Section, we reviewed the related work. In Sects. 3, 4, and 5, we presented solutions for the above goals, respectively. Finally, Sect. 6 gives some concluding remarks.

2   Related Work

Data mining can be used to design an analysis process for exploiting big agricultural datasets. Recently, many papers have been published that exploit machine learning algorithms on sensor data and build models to improve agricultural economics, such as [23–25]. In these, the paper [23] predicted crop yield by using self-organizing-maps supervised learning models; namely supervised Kohonen networks, counter-propagation artificial networks and XY-fusion. The paper [24] predicted drought conditions by using three rule-based machine learning; namely random forest, boosted regression trees, and Cubist. Finally, the paper [25] predicted pest population dynamics by using time series clustering and structural change detection which detected groups of different pest species. However, the proposed solutions are not satisfied the problems of agricultural Big Data, such as data integration, data schema, storage capacity, security and performance.

From a Big Data point of view, the papers [14] and [26] have proposed "smart agricultural frameworks". In [14], the platform used Hive to store and analyse sensor data about land, water and biodiversity which can help increase food production with lower environmental impact. In [26], the authors moved toward a notion of climate analytics-as-a-service by building a high-performance analytics and scalable data management platform which is based on modern infrastructures, such as Amazon web services, Hadoop and Cloudera. However, the two papers did not discuss how to build and implement a DW for a precision agriculture.

Our approach is inspired by papers [20,27,28] and [19] which presented ways of building a DW for agricultural data. In [28], the authors extended entity-relationship model for modelling operational and analytical data which is called the multi-dimensional entity-relationship model. They introduced new representation elements and showed the extension of an analytical schema. In [27], a relational database and an RDF triple store, were proposed to model the overall datasets. In that, the data are loaded into the DW in RDF format, and cached in the RDF triple store before being transformed into relational format. The actual data used for analysis was contained in the relational database. However, as the schemas in [28] and [27] were based on entity-relationship models, they cannot deal with high-performance, which is the key feature of a data warehouse.

In [20], a star schema model was used. All data marts created by the star schemas are connected via some common dimension tables. However, a star schema is not enough to present complex agricultural information and it is difficult to create new data marts for data analytics. The number of dimensions of DW proposed by [20] is very small; only 3-dimensions – namely, Species, Location, and Time. Moreover, the DW concerns livestock farming. Overcoming disadvantages of the star schema, the paper [19] proposed a constellation schema for an agricultural DW architecture in order to facilitate quality criteria of a DW. However, it did not describe how to implement the proposed DW. Finally, all papers [19,20,27,28] did not used Hive, MongoDB or Cassandra in their proposed DWs.

3   Analyzing Cassandra, MongoDB and Hive in Agricultural Big Data

In general, a DW is a federated repository for all the data that an enterprise can collect through multiple heterogeneous data sources belonging to various enterprise's business systems or external inputs [9,13]. A quality DW should adapt many important criteria [1,15], such as: (1) Making information easily accessible; (2) Presenting and providing right information at the right time; (3) Integrating data and adapting to change; (4) Achieving tangible and intangible benefits; (5) Being a secure bastion that protects the information assets; and (6) Being accepted by DW users. So, to build an efficient agricultural DW, we need to take into account these criteria.

Currently, there are many popular databases that support efficient DWs, such as such as Redshift, Mesa, Cassandra, MongoDB and Hive. Hence, we are analyzing the most popular and see which is the best suited for our data problem. In these databases, Redshift is a fully managed, petabyte-scale DW service in the cloud which is part of the larger cloud-computing platform Amazon Web Services [2]. Mesa is highly scalable, petabyte data warehousing system which is designed to satisfy a complex and challenging set of users and systems requirements related to Google's Internet advertising business [10]. However, Redshift and Mesa are not open source. While, Cassandra, MongoDB and Hive are open source databases, we want to use them to implement agriculture DW. Henceforth, the Cassandra and MongoDB terms are used to refer to DWs of Cassandra and MongoDB databases.

There are many papers studying Cassandra, MongoDB and Hive in the view of general DWs. In the following two subsections, we present advantages, disadvantages, similarities and differences between Cassandra, MongoDB and Hive in the context of agricultural DW. Specially, we analyze to find how to combine these DWs together to build a DW for agricultural Big Data, not necessarily best DW.

3.1   Advantages and Disadvantages

Cassandra, MongoDB and Hive are used widely for enterprise DWs. Cassandra[1] is a distributed, wide-column oriented DW from Apache that is highly scalable and designed to handle very large amounts of structured data. It provides high availability with no single point of failure, tuneable and consistent. Cassandra offers robust support for transactions and flexible data storage based on ideas of DynamoDB and BigTable [11,18]. While, MongoDB[2] is a powerful, cross-platform, document oriented DW that provides, high performance, high availability, and scalability [4,12]. It works on concept of collection and document, JSON-like documents, with dynamic schemas. So, documents and data structure can be changed over time. Secondly, MongoDB combines the ability to scale out with features, such as ad-hoc query, full-text search and secondary index. This provides powerful ways to access and analyze datasets.

Hive[3] is an SQL data warehouse infrastructure on top of Hadoop[4] for writing and running distributed applications to summarize Big Data [5,16]. Hive can

---

[1] http://cassandra.apache.org.
[2] http://mongodb.com.
[3] http://hive.apache.org.
[4] http://hadoop.apache.org.

be used as an online analytical processing (OLAP) system and provides tools to enable data extract - transform - load (ETL). Hive's metadata structure provides a high-level, table-like structure on top of HDFS (Hadoop Distributed File System). That will significantly reduce the time to perform semantic checks during the query execution. Moreover, by using Hive Query Language (HQL), similar to SQL, users can make simple queries and analyse the data easily.

Although, the three DWs have many advantages and have been used widely, they have major limitations. These limitations impact heavily on their use as agricultural DW.

1. In Cassandra: (1) Query Language (CQL) does not support joint and sub-query, and has limited support for aggregations that are difficult to analyze data; (2) Ordering is done per-partition and specified at table creation time. The sorting of thousands or millions of rows can be fast in development but sorting billion ones is a bad idea; (3) A single column value is recommended not be larger than 1 MB that is difficult to contain videos or high quality images, such as LiDAR images, 3-D images and satellite images.
2. In MongoDB: (1) The maximum BSON document size is 16 MB that is difficult to contain large data such as video, audio and high quality image; (2) JSON's expressive capabilities are limited because the only types are null, boolean, numeric, string, array, and object; (3) We cannot automatically roll-back more than 300 MB of data. If we have more than that, manual intervention is needed.
3. Hive is not designed for: (1) Online transaction processing; (2) Real-time queries; (3) Large data on network; (4) Trivial operations; (5) Row-level update; and (6) Iterative execution.

3.2   Feature Comparison

Table 1 lists technical features used to compare Hive, MongoDB and Cassandra. For the ten overview features given in section A of Table 1, the three DWs differ in data schema, query language and access methods. However, they all support map reduce. Moreover, the ETL feature is supported by Hive, limited to Cassandra and unsupported by MongoDB. The full-text search feature is only supported by MongoDB. The secondary index and ad-hoc query features are supported by Hive and MongoDB but not or restricted by Cassandra. The 9th feature being the Consistency – Availability – Partition tolerant classification (CAP) theorem says how the database system behaves when facing network instability. It implies that in the presence of a network partition, one has to choose between consistency and availability. Hive and Cassandra choose availability. While, MongoDB chooses consistency. Finally, the structure of Hive and MongoDB are master - slave while Cassandra has peer - to - peer structure.

The section B of Table 1 describes five industrial features, such as governance, monitoring, data lifecycle management, workload management, and replication-recovery. All of Hive, MongoDB and Cassandra support these features. Hive supports governance and data lifecycle management features via Hadoop. Cassandra is based on Java Management Extensions (JME) for governance.

Table 1. Technical features

No.	Features	Hive	MongoDB	Cassandra
A. Overview Features				
1	Data scheme	Yes	No-Schema	Flexible Schema
2	Query language	HQL	JS-like syntax	CQL
3	Accessing method	JDBC, ODBC, Thrift	JSON	Thrift
4	Map reduce	Yes	Yes	Yes
5	ETL	Yes	No	Limited
6	Full-text search	No	Yes	No
7	Ad-hoc query	Yes	Yes	No
8	Secondary index	Yes	Yes	Restricted
9	CAP	AP	CP	AP
10	Structure	Master – Slave	Master – Slave	Peer – to – Peer
B. Industrial Features				
1	Governance	Yes (via Hadoop)	Yes	Yes (via JME)
2	Monitoring	Yes	Yes	Yes
3	Data lifecycle management	Yes (via Hadoop)	Yes	Yes
4	Workload management	Yes	Yes	Yes
5	Replication-Recovery	Yes	Yes	Yes

The data management and DW features are described in section A and section B of Table 2, respectively. The data management features are security, high storage capacity, and data ingestion and pre-processing. The DWs have support for these features, except Cassandra does not support for data ingestion and pre-processing. Hive has the best for high storage capacity. The DW features are business intelligent, data science and high performance. Hive supports well business intelligent and data science but it is not suitable for real-time performance. MongoDB is very fast but it is limited in supporting for business intelligent and data science. Cassandra also is very fast and supports business intelligent but has limited capabilities for data science.

Table 2. Data Management and Data Warehouse Features

No.	Features	Hive	MongoDB	Cassandra
A. Data Management Features				
1	Security	Yes	Yes	Yes
2	High storage capacity	Yes (best)	Yes	Yes
3	Data ingestion and pre-processing	Yes	Yes	No
B. Data Warehouse Features				
1	Business intelligent	Very good	Limited	Good
2	Data science	Very good	Limited	Limited
3	High performance	Non-real time	Real time	Real time

4   Agricultural Data Warehouse

The general architecture of a typical DW includes four separate and distinct modules being Raw Data, ETL, Integrated Information and Data Mining. In the scope of this paper, we focus on the Integrated Information module which is a logically a centralised repository. It includes DW storage, data marts, data cubes and OLAP engine.

The DW storage is organised, stored and accessed using a suitable schema defined in the metadata. It can be either directly accessed or used to creating data marts which is usually oriented to a particular business function or enterprise department. A data cube is a data structure that allows fast analysis of data according to the multiple dimensions that define a business problem. The data cubes are created by the OLAP engine.

4.1   OLAP

OLAP is a category of software technology that provides the insight and understanding of data in multiple dimensions through fast, consistent, interactive access to enable analysts or managers to make better decisions. By using roll-up, drill-down, slice-dice and pivot operations, OLAP performs multidimensional analysis in a wide variety of possible views of information that provide complex calculations, trend analysis and sophisticated data modelling with a short execution time. So, OLAP is a key way to exploit information in a DW to allow end-users to analyze and explore data in multidimensional views.

The OLAP systems are categorised into three types: namely relational OLAP (ROLAP), multidimensional OLAP (MOLAP) and hybrid OLAP (HOLAP). In our agricultural Big Data context, HOLAP is more suitable than ROLAP and MOLAP because:

1. ROLAP has quite slow performance. Each ROLAP report is an SQL query in the relational database that requires a significant execution time. In addition, ROLAP does not meet all the users' needs, especially complex queries.
2. MOLAP requires that all calculations should be performed during the data cube construction. So, it handles only a limited amount of data and does not scale well. In addition, MOLAP is not capable of handling detailed data.
3. HOLAP inherits relational technique of ROLAP to store large data volumes and detailed information. Additionally, HOLAP also inherits multidimensional techniques of MOLAP to perform complex calculations and has good performance.

4.2   The Proposed Architecture

Based on the analyis in Sect. 3, Hive is chosen for building our DW storage and it is combining with MongoDB to implement our Integrated Information module. So, Hive contains database created from the our DW schema in the initialization period. This is for the following reasons:

1. Hive is based on Hadoop which is the most powerful tool of Big Data. Besides, HQL is similar to SQL which is familiar to the majority of users. Especially, Hive supports well high storage capacity, business intelligent and data science more than MongoDB and Cassandra. These features of Hive are useful to make an agricultural DW and apply data mining technologies.
2. Hive does not have real-time performance so it needs to be combined with MongoDB or Cassandra to improve performance of our Integrated Information module.
3. MongoDB is more suitable than Cassandra to complement Hive because: (1) MongoDB supports joint operation, full text search, ad-hoc query and second index which are helpful to interact with users. While Cassandra does not support these features; (2) MongoDB has the same master – slave structure with Hive that is easy to combine. While the structure of Cassandra is peer - to - peer; (3) Hive and MongoDB are more reliable and consistent. So the combination between Hive and MongoDB supports fully the CAP theorem while Hive and Cassandra are the same AP systems.

Our DW architecture for agricultural Big Data is illustrated in Fig. 1 which contains three modules, namely Integrated Information, Products and Raw Data. The Integrated Information module includes two components being MongoDB component and Hive component. Firstly, the MongoDB component will receive real-time data, such as user data, logs, sensor data or queries from Products module, such as web application, web portal or mobile app. Besides, some results which need to be obtained in real-time will be transferred from the MongoDB to Products. Second, the Hive component will store the online data from and send the processed data to the MongoDB module. Some kinds of queries having

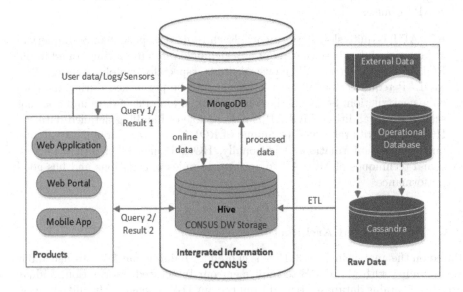

Fig. 1. Our agricultural data warehouse architecture

complex calculations will be sent directly to Hive. After that, Hive will send the results directly to the Products module.

In Raw Data module, almost data in Operational Databases or External Data components is loaded into Cassandra component. It means that we use Cassandra to represent raw data storage. In the idle times of the system, the update raw data in Cassandra will be imported into Hive through the ELT tool. This improves the performance of ETL and helps us deploy our system on cloud or distributed systems better.

4.3 Our Schema

The DW uses schema to logically describe the entire datasets. A schema is a collection of objects, including tables, views, indexes, and synonyms which consist of some fact and dimension tables [21]. The DW schema can be designed through the model of source data and the requirements of users. There are three kind of

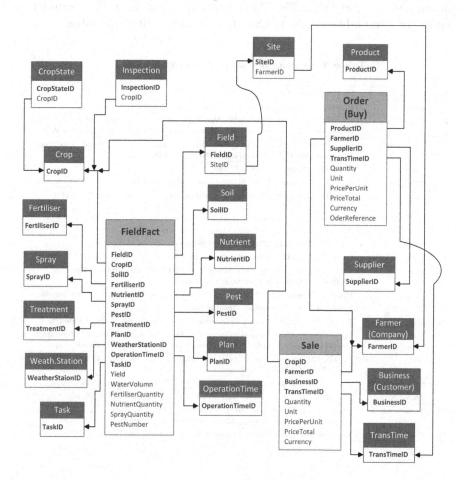

Fig. 2. A part of our data warehouse schema for Precision Agriculture

schemas, namely star, snowflake and constellation. With features of agricultural data, the agricultural DW schema needs to have more than one fact table and be flexible. So, the constellation schema, also known galaxy schema, is selected to design our DW schema.

We developed a constellation schema for our agricultural DW and it is partially described in Fig. 2. It includes 3 fact tables and 19 dimension tables. The FieldFact fact table contains data about agricultural operations on fields. The Order and Sale fact tables contain data about farmers' trading operations. The FieldFact, Order and Sale facts have 12, 4 and 4 dimensions, and have 6, 6 and 5 measures, respectively. While, dimension tables contain details about each instance of an object involved in a crop yield.

The main attributes of these dimension tables are described in the Table 3. The key dimension tables are connected to their fact table. However, there are some dimension tables connected to more than one fact table, such as Crop and Farmer. Besides, the CropState, Inspection and Site dimension tables are not connected to any fact table. The CropState and Inspection tables are used to support the Crop table. While, the Site table supports the Field table.

Table 3. Descriptions of some dimension tables

No.	Dim. tables	Particular attributes
1	Business	BusinessID, Name, Address, Phone, Mobile, Email
2	Crop	CropID, CropName, VarietyID, VarietyName, EstYield, SeasontSart, SeasonEnd, BbchScale, ScientificName, HarvestEquipment, EquipmentWeight
3	CropState	CropStateID, CropID, StageScale, Height, MajorStage, MinStage, MaxStage, Diameter, MinHeight, MaxHeight, CropCoveragePercent
4	Farmer	FarmerID, Name, Address, Phone, Mobile, Email
5	Fertiliser	FertiliserID, Name, Unit, Status, Description, GroupName
6	Field	FieldID, Name, SiteID, Reference, Block, Area, WorkingArea, FieldGPS, Notes
7	Inspection	InspectionID, CropID, Description, ProblemType, Severity, ProblemNotes, AreaValue, AreaUnit, Order, Date, Notes, GrowthStage
8	Nutrient	NutrientID, NutrientName, Date, Quantity
9	OperationTime	OperationTimeID, StartDate, EndDate, Season
10	Pest	PestID, CommonName, ScientificName, PestType, Description, Density, MinStage, MaxStage, Coverage, CoverageUnit
11	Plan	PlanID, PName, RegisNo, ProductName, ProductRate, Date, WaterVolume
12	Product	ProductID, ProductName, GroupName
13	Site	SiteID, FarmerID, SName, Reference, Country, Address, GPS, CreatedBy
14	Spray	SprayID, SProductName, ProductRate, Area, Date, WaterVol, ConfDuration, ConfWindSPeed, ConfDirection, ConfTemp, ConfHumidity, ActivityType
15	Soil	SoilID, PH, Phosphorus, Potassium, Magnesium, Calcium, CEC, Silt, Clay, Sand, TextureLabel, TestDate
16	Supplier	SupplierID, Name, ContactName, Address, Phone, Mobile, Email
17	Task	TaskID, Desc, Status, TaskDate, TaskInterval, CompDate, AppCode
18	Treatment	TreatmentID, TreatmentName, FormType, LotCode, Rate, ApplCode, LevlNo, Type, Description, ApplDesc, TreatmentComment
19	WeatherStation	WeatherStationID, Name, MeasureDate, AirTemp, SoilTemp, WindSpeed

5   Experiments

Through the proposed architecture in Sect. 4.2, our DW inherited many advantages from Hive, MongoDB and Cassandra presented in Sect. 3, such as high performance, high storage, large scale analytic and security. In the scope of this paper, we evaluated our DW schema and data analysis capacity on real agricultural datasets through complex queries. In addition, the time performance of our agricultural DW storage was also evaluated and compared to MySQL on many particular built queries belonging to different query groups.

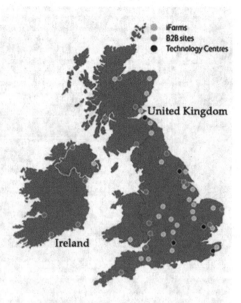

Fig. 3. Data in UK and Ireland [22]

5.1   Data Analyzing Demo

The input data for the DW was primarily obtained from an agronomy company which supplies data from its operational systems, research results and field trials. Specially, we are supplied real agricultural data in iFarms, B2B sites, technology centres and demonstration farms. Their specific positions in several European countries are presented in Figs. 3 and 4 [22]. There is a total of 29 datasets. On average, each dataset contains 18 tables and is about 1.4 GB in size. The source datasets are loaded on our CONSUS DW Storage based on the schema described in Sect. 4.3 through an ETL tool. From the DW storage, we can extract and analyze useful information through tasks using complex HQL queries or data mining algorithms. These tasks could not be executed if the separate 29 datasets have not been integrated into our DW storage.

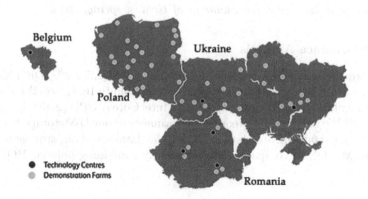

Fig. 4. Data in Continental Europe [22]

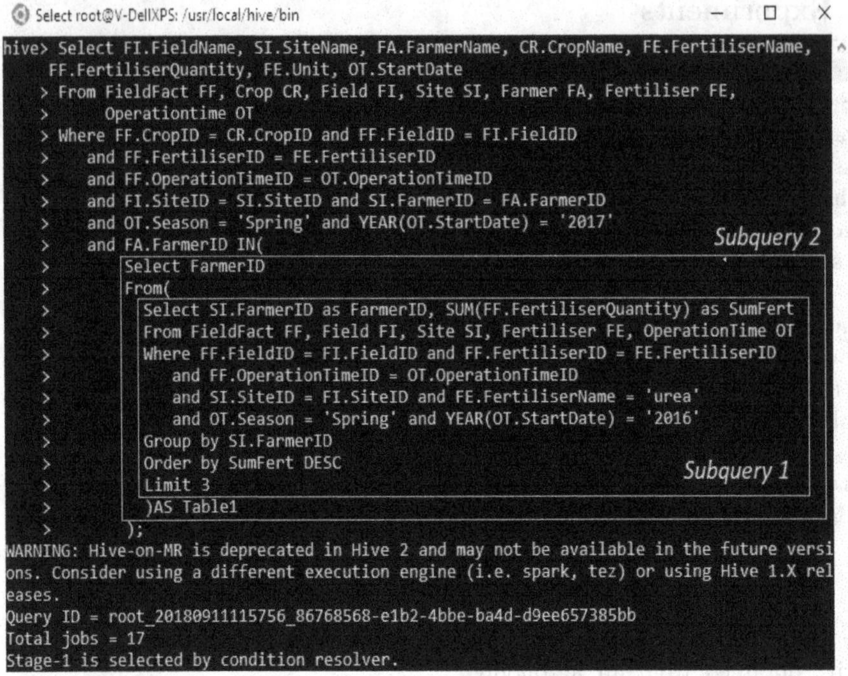

```
Select root@V-DellXPS: /usr/local/hive/bin                                    —  □  X

hive> Select FI.FieldName, SI.SiteName, FA.FarmerName, CR.CropName, FE.FertiliserName,
    FF.FertiliserQuantity, FE.Unit, OT.StartDate
  > From FieldFact FF, Crop CR, Field FI, Site SI, Farmer FA, Fertiliser FE,
  >     Operationtime OT
  > Where FF.CropID = CR.CropID and FF.FieldID = FI.FieldID
  >     and FF.FertiliserID = FE.FertiliserID
  >     and FF.OperationTimeID = OT.OperationTimeID
  >     and FI.SiteID = SI.SiteID and SI.FarmerID = FA.FarmerID
  >     and OT.Season = 'Spring' and YEAR(OT.StartDate) = '2017'
  >     and FA.FarmerID IN(                                         Subquery 2
  >         Select FarmerID
  >         From(
  >             Select SI.FarmerID as FarmerID, SUM(FF.FertiliserQuantity) as SumFert
  >             From FieldFact FF, Field FI, Site SI, Fertiliser FE, OperationTime OT
  >             Where FF.FieldID = FI.FieldID and FF.FertiliserID = FE.FertiliserID
  >                 and FF.OperationTimeID = OT.OperationTimeID
  >                 and SI.SiteID = FI.SiteID and FE.FertiliserName = 'urea'
  >                 and OT.Season = 'Spring' and YEAR(OT.StartDate) = '2016'
  >             Group by SI.FarmerID
  >             Order by SumFert DESC
  >             Limit 3                                             Subquery 1
  >         )AS Table1
  >     );
WARNING: Hive-on-MR is deprecated in Hive 2 and may not be available in the future versi
ons. Consider using a different execution engine (i.e. spark, tez) or using Hive 1.X rel
eases.
Query ID = root_20180911115756_86768568-e1b2-4bbe-ba4d-d9ee657385bb
Total jobs = 17
Stage-1 is selected by condition resolver.
```

Fig. 5. A screenshort of executing the query example in our Hive

An example for a complex request: *"List crops, fertilisers, corresponding fertiliser quantities in spring, 2017 in every field and site of 3 farmers (crop companies) who used the large amount of Urea in spring, 2016"*. In our schema, this query can be executed by a HQL/SQL query as shown in Fig. 5. To execute this request, the query needs to exploit data in the FieldFact fact table and the six dimension tables, namely Crop, Field, Site, Farmer, Fertiliser and OperationTime. The query consists of two subqueries which return *3 farmers (crop companies) that used the largest amount of Urea in spring, 2016*.

5.2 Performance Analysis

The performance analysis was implemented using MySQL 5.7.22, JDK 1.8.0_171, Hadoop 2.6.5 and Hive 2.3.3 which run on Bash on Ubuntu 16.04.2 on Windows 10. All experiments were run on a laptop with an Intel Core i7 CPU (2.40 GHz) and 16 GB memory. We only evaluate reading performance of our DW storage because a DW is used for reporting and data analysis. The database of our storage is duplicated into MySQL to compare performance. By combining popular HQL/SQL commands, namely Where, Group by, Having, Left (right) Join, Union and Order by, we create 10 groups for testing. Every group has 5 queries and uses one, two or more commands (see Table 4). Besides, every query also uses operations, such as And, Or, $\geq$, Like, Max, Sum and Count, to combine with the commands.

All queries were executed three times and we took the average value of the these executions. The different times in runtime between MySQL and our storage of query q_i is calculated as $Times_{q_i} = RT_{q_i}^{mysql}/RT_{q_i}^{ours}$. Where, $RT_{q_i}^{mysql}$ and $RT_{q_i}^{ours}$ are respectively average runtimes of query q_i on MySQL and our storage. Besides, with each group G_i, the different times in runtime between MySQL and our storage $Times_{G_i} = RT_{G_i}^{mysql}/RT_{G_i}^{ours}$. Where, $RT_{G_i} = Average(RT_{q_i})$ is average runtime of group G_i on MySQL or our storage.

Table 4. Command combinations of queries

Group	Queries	Where	Group by	Having	Left (right) Joint	Union	Order by
1	1–5	x					
2	6–10	x	x				
3	11–15	x			x		
4	16–20	x				x	
5	21–25	x					x
6	26–30	x			x		x
7	31–35	x	x	x			
8	36–40	x	x	x			x
9	41–45	x	x	x	x		x
10	45–50	x	x	x		x	x

Figure 6 describes different times between MySQL and our storage in runtime of every query belongs to 10 groups. Unsurprisingly, although running on one computer, but with large data volume, our storage is faster than MySQL at 46/50 queries and all 10 query groups. MySQL is faster than our storage at 3 queries 12th, 13th and 18th belonging to groups 3rd and 4th. Two databases are same at the query 25th belonging to group 5th. Within each query group, to have a fair performance comparison, the queries combine randomly fact tables and dimensional tables. This makes the complex of queries having far high difference. Combining with different size and structure of the tables, it make the runtime of queries being huge differences although belonging a group, as presented in Fig. 6.

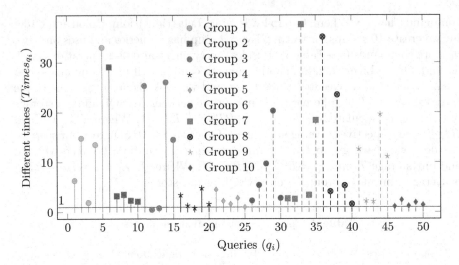

Fig. 6. Different times between MySQL and our storage in runtime of every Query

Beside comparing runtime in every query, we aslo compare runtime of every group presented in Fig. 7. Comparing to MySQL, our storage is more than at most (6.24 times) at group 1st which uses only *Where* command, and at least (1.22 times) at group 3rd which uses *Where* and *Joint* commands.

Figure 8 presents the average runtime of the 10 query groups on MySQL and our storage. Mean, the run time of a reading query on MySQL and our storage is 687.8 s and 216.1 s, respectively. It means that

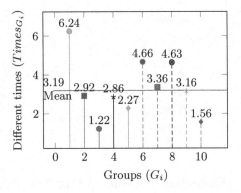

Fig. 7. Different times between MySQL and our storage in runtime of every group

our storage is faster 3.19 times. In the future, by deploying our storage solution on cloud or distributed systems, we believe that the performance will be even much better than MySQL.

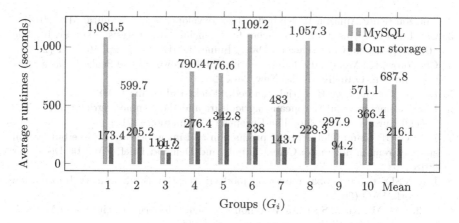

Fig. 8. Average Runtimes of MySQL and our storage in every Groups

6   Conclusion and Future Work

In this paper, we compared and analyzed some existing popular open source DWs in the context of agricultural Big Data. We designed and implemented the agricultural DW by combining Hive, MongoDB and Cassandra DWs to exploit their advantages and overcome their limitations. Our DW includes necessary modules to deal with large scale and efficient analytics for agricultural Big Data. Additionally, the presented schema herein was optimised for the real agricultural datasets that were made available to us. The schema been designed as a constellation so it is flexible to adapt to other agricultural datasets and quality criteria of agricultural Big Data. Moreover, using the short demo, we outlined a complex HQL query that enabled knowledge extraction from our DW to optimize of agricultural operations. Finally, through particular reading queries using popular HQL/SQL commands, our DW storage outperforms MySQL by far.

In the future work, we shall pursue the deployment of our agricultural DW on a cloud system and implement more functionalities to exploit this DW. The future developments will include: (1) Sophisticated data mining techniques [3] to determine crop data characteristics and combine with expected outputs to extract useful knowledge; (2) Predictive models based on machine learning algorithms; (3) An intelligent interface for data access; (4) Combination with the high-performance knowledge map framework [17].

Acknowledgment. This research is part of the CONSUS research programme which is funded under the SFI Strategic Partnerships Programme (16/SPP/3296) and is co-funded by Origin Enterprises Plc.

References

1. Adelman, S., Moss, L.: Data Warehouse Project Management, 1st edn. Addison-Wesley Professional, Boston (2000)

2. Amazon team: Amazon Redshift database developer guide. Samurai ML (2018)
3. Cai, F., et al.: Clustering approaches for financial data analysis: a survey. In: The 8th International Conference on Data Mining (DMIN 2012), pp. 105–111 (2012)
4. Chodorow, K.: MongoDB: The Definitive Guide. Powerful and Scalable Data Storage, 2nd edn. O'Reilly Media, New York (2013)
5. Du, D.: Apache Hive Essentials, 2nd edn. Packt Publishing (2018)
6. Eurobarometer team: Europeans, agriculture and the common agricultural policy. Special Eurobarometer 440, The European Commission (2016)
7. FAO-CSDB team: Global cereal production and inventories to decline but overall supplies remain adequate. Cereal Supply and Demand Brief, FAO, 06 December 2018
8. FAO-FSIN team: Global report on food crises 2018. Food Security Information Network, FAO (2018)
9. Golfarelli, M., Rizzi, S.: Data Warehouse Design: Modern Principles and Methodologies. McGraw-Hill Education, New York (2009)
10. Gupta, A., et al.: Mesa: a geo-replicated online data warehouse for Google's advertising system. Commun. ACM **59**(7), 117–125 (2016)
11. Hewitt, E., Carpenter, J.: Cassandra: The Definitive Guide. Distributed Data at Web Scale, 2nd edn. O'Reilly Media, New York (2016)
12. Hows, D., et al.: The Definitive Guide to MongoDB. A Complete Guide to Dealing with Big Data Using MongoDB, 3rd edn. Apress, Berkely (2015)
13. Inmon, W.H.: Building the Data Warehouse. Wiley, New York (2005)
14. Kamilaris, A., et al.: Estimating the environmental impact of agriculture by means of geospatial and big data analysis. Science to Society, pp. 39–48 (2018)
15. Kimball, R., Ross, M.: The Data Warehouse Toolkit: The Definitive Guide to Dimensional Modeling, 3rd edn. Wiley, New York (2013)
16. Lam, C.P., et al.: Hadoop in Action, 2nd edn. Manning, Greenwich (2016)
17. Le-Khac, N.-A., et al.: Distributed knowledge map for mining data on grid platforms. Int. J. Comput. Sci. Network Secur. **7**(10), 98–107 (2007)
18. Neeraj, N.: Mastering Apache Cassandra, 2nd edn. Packt Publishing, Birmingham (2015)
19. Ngo, V.M., et al.: An efficient data warehouse for crop yield prediction. In: The 14th International Conference Precision Agriculture (ICPA-2018), pp. 3:1–3:12 (2018)
20. Nilakanta, S., et al.: Dimensional issues in agricultural data warehouse designs. Comput. Electron. Agric. **60**(2), 263–278 (2008)
21. Oracle team: Database data warehousing guide, Oracle12c doc release 1 (2017)
22. Origin team: Annual report and accounts, Origin Enterprises plc (2018)
23. Pantazi, X.E.: Wheat yield prediction using machine learning and advanced sensing techniques. Comput. Electron. Agric. **121**, 57–65 (2016)
24. Park, S., et al.: Drought assessment and monitoring through blending of multi-sensor indices using machine learning approaches for different climate regions. Agric. Forest Meteorol. **216**, 157–169 (2016)
25. Rupnik, R., et al.: AgroDSS: a decision support system for agriculture and farming. Computers and Electronics in Agriculture (2018)
26. Schnase, J., et al.: MERRA analytic services: meeting the big data challenges of climate science through cloud-enabled climate analytics-as-a-service. Comput. Environ. Urban Syst. **61**, 198–211 (2017)
27. Schuetz, C.G., et al.: Building an active semantic data warehouse for precision dairy farming. Organ. Comput. Electron. Commerce **28**(2), 122–141 (2018)

28. Schulze, C., et al.: Data modelling for precision dairy farming within the competitive field of operational and analytical tasks. Comput. Electron. Agric. **59**(1–2), 39–55 (2007)
29. UN team: World population projected to reach 9.8 billion in 2050, and 11.2 billion in 2100. Department of Economic and Social Affairs, United Nations (2017)
30. USDA report: World agricultural supply and demand estimates 08/2018. United States Department of Agriculture (2018)

FFD: A Federated Learning Based Method for Credit Card Fraud Detection

Wensi Yang[1,2], Yuhang Zhang[1,2], Kejiang Ye[1], Li Li[1],
and Cheng-Zhong Xu[3(✉)]

[1] Shengzhen Institutes of Advanced Technology, Chinese Academy of Sciences,
Shenzhen 518055, China
[2] University of Chinese Academy of Sciences, Beijing 100049, China
[3] Department of Computer and Information Science,
Faculty of Science and Technology, State Key Laboratory of IoT for Smart City,
University of Macau, Taipa, Macao, Special Administrative Region of China
cz.xu@siat.ac.cn

Abstract. Credit card fraud has caused a huge loss to both banks and consumers in recent years. Thus, an effective Fraud Detection System (FDS) is important to minimize the loss for banks and cardholders. Based on our analysis, the credit card transaction dataset is very skewed, there are much fewer samples of frauds than legitimate transactions. Furthermore, due to the data security and privacy, different banks are usually not allowed to share their transaction datasets. These problems make FDS difficult to learn the patterns of frauds and also difficult to detect them. In this paper, we propose a framework to train a fraud detection model using behavior features with federated learning, we term this detection framework FFD (Federated learning for Fraud Detection). Different from the traditional FDS trained with data centralized in the cloud, FFD enables banks to learn fraud detection model with the training data distributed on their own local database. Then, a shared FDS is constructed by aggregating locally-computed updates of fraud detection model. Banks can collectively reap the benefits of shared model without sharing the dataset and protect the sensitive information of cardholders. Furthermore, an oversampling approach is combined to balance the skewed dataset. We evaluate the performance of our credit card FDS with FFD framework on a large scale dataset of real-world credit card transactions. Experimental results show that the federated learning based FDS achieves an average of test AUC to 95.5%, which is about 10% higher than traditional FDS.

Keywords: Federated learning · Credit card fraud · Skewed dataset

1 Introduction

Credit card transactions take place frequently with the improvement of modern computing technology and global communication. At the same time, fraud is also increasing dramatically. According to the European Central Bank report [1], billions of Euros are lost in Europe because of credit card fraud every year.

© Springer Nature Switzerland AG 2019
K. Chen et al. (Eds.): BigData 2019, LNCS 11514, pp. 18–32, 2019.
https://doi.org/10.1007/978-3-030-23551-2_2

Credit card is considered as a nice target of fraud since a significant amount of money can be obtained in a short period with low risk [2]. Credit card frauds can be made in different forms, such as application fraud [3], counterfeit cards [4], offline fraud and online fraud [5]. Application fraud is a popular and dangerous fraud, it refers that fraudsters acquire a credit card by using false personal information or other person's information with the intention of never repaying the purchases [3]. Counterfeit fraud occurs when the credit card is used remotely; only the credit card details are needed [6]. Offline fraud happens when the plastic card was stolen by fraudsters, using it in stores as the actual owner while online fraud is committed via web, phone shopping or cardholder not-present [5].

There are two mechanisms that are widely used to combat fraud – fraud prevention and fraud detection. Fraud prevention, as the first line of defense, is to filter high risk transactions and stop them occurring at the first time. There are numerous authorization techniques for credit card fraud prevention, such as signatures [7], credit card number, identification number, cardholder's address and expiry data, etc. However, these methods are inconvenient for the customers and are not enough to curb incidents of credit card fraud. There is an urgent need to use fraud detection approaches which analyze data that can detect and eliminate credit card fraud [8].

However, there are many constraints and challenges that hinder the development of an ideal fraud detection system for banks. Existing FDS usually is prone to inefficient, with a low accuracy rate, or raises many false alarm, due to the reasons such as dataset insufficiency, skewed distribution and limitation of detection time.

- **Dataset Insufficiency**
  One of the main issues associated with the FDS is the lack of available public datasets [9]. The increasing concern over data privacy imposes barriers to data sharing for banks. At the same time, most fraud detection systems are produced in-house concealing the model details to protect data security. However, a reliable credit card FDS is impossible to be established in the absence of available dataset.
- **Skewed Distribution**
  Credit card transactions are highly unbalanced in every bank - where a few samples are fraud while a majority of them are legitimate transactions. In most circumstance, 99% of transactions are normal while fraudulent transactions are less than 1% [10]. In this case, it is very difficult for machine learning algorithms to discover the patterns in the minority class data. Furthermore, skewed class distribution has a serious impact on the performance of classifiers that are tend to be overwhelmed by the majority class and ignore the minority class [11].
- **Limitation of Detection Time**
  In some online credit card payment applications, the delay in time can lead to intolerable loss or potential exploitation by fraudsters. Therefore, an online FDS that has the ability to deal with limited time resource and qualifies enough to detect fraudulent activities rapidly is extremely important [12]. Building a good fraud detection framework which is fast enough to be utilized in a real-time environment should be considered.

In this paper, we aim to address these issues with a novel fraud detection system. First, we focus on a fraud detection system which can protect the data privacy, meanwhile, it can be shared with different banks. Then, we solve the problem of skewed distribution of datasets. A federated fraud detection framework with data balance approach is proposed to construct a fraud detection model, which is different from previous FDS. Federated fraud detection framework enables different banks to collaboratively learn a shared model while keeping all the training data which is skewed on their own private database. Furthermore, the accuracy, convergence rate, training time and communication cost of FDS are comprehensively taken into consideration.

The main contributions of this paper are summarized as follows:

(1) To deal with fraud detection problem and construct an effective FDS in data insufficient circumstance. A kind of decentralized data machine learning algorithm–federated fraud detection framework is proposed to train fraud detection model with the fraud and legitimate behavior features. Our work takes a step forward by developing ideas that solve the problem of dataset insufficiency for credit card FDS.
(2) Using the real-world dataset from the European cardholders, experiments are conducted to demonstrate our method is robust to the unbalanced credit card data distributions. Experimental results depicted that credit card FDS with federated learning improves traditional FDS[1] by 10% AUC and F1 score.
(3) From the results of experiments, conclusions that how to coordinate communication cost and accuracy of FDS are made, which would be helpful for making a trade off between computation resources and real-time FDS for future fraud detection work.

The rest of the paper is organized as follows. In Sect. 2, related work about credit card fraud is discussed. Section 3 gives the details of federated fraud detection framework. Section 4 provides an analysis of the dataset and experimental results. Conclusions and future work are presented in Sect. 5.

2   Related Work

Although fraud detection in the credit card industry is a much-discussed topic which receives a lot of attention, the number of public available works is rather limited [14]. One of the reasons is that credit card issuers protect the sharing of data source from revealing cardholder's privacy. In literature about credit card fraud detection, the data mining technologies used to create credit card FDS can categorized into two types: supervised method and unsupervised method.

Supervised learning techniques relies on the dataset that has been labeled as 'normal' and 'fraud'. This is the most prevalent approach for fraud detection. Recently, decision tree combined with contextual bandits are proposed to

---

[1] The traditional ensemble FDS [13] with SMOTE(borderline2) balancing techniques achieved AUC of 88% and F1 score of 82% on the same dataset.

construct a dynamic credit card fraud detection model [15]. Adaptive learning algorithms which can update fraud detection model for streaming evolving data over time [16] to adapt with and capture changes of patterns of fraud transactions. Data level balanced techniques such as under sampling approach [17], SMOTE and EasyEnsemble are conducted in [18] to find out the most efficient mechanism for credit card fraud detection. A supervised ensemble method [19] was developed by combining the bagging and boosting techniques together. Bagging technique used to reduce the variance for classification model through resampling the original data set, while boosting technique reduce the bias of the model for unbalanced data. A FDS constructed with a scalable algorithm BOAT (Boostrapped Optimistic Algorithm for Tree Construction) which supports several levels of the tree in one scan over the training database to reduce training time [8]. Other supervised learning methods in fraud are Bayes [20], artificial neural network(ANN) [21,22] and support vector machine [23,24].

In unsupervised learning, there is no class label for fraud detection model construction. As in [25], it proposed unsupervised methods that do not require the accurate label of fraudulent transactions but instead detect changes in behavior or unusual transactions. K-means clustering algorithm is an unsupervised learning algorithm for grouping a given set of data based on the similarity in their attribute used to detect credit card fraud [26].

The advantages of supervised FDS over semi-supervised and unsupervised FDS is that the outputs manipulated by supervised FDS are meaningful to humans, and it can be easily used for discriminative patterns. In this paper, a data level balance approach – SMOTE is used to handle the problem of skewed distribution by oversampling fruad transactions. Supervised method with a deep network (CNN) is applied by participated banks to detect fraud transactions. Federated fraud detection framework balances the FDS performance and training time by controlling deep network learning process. But one of the biggest differences is that fraud detection models described above are only trained by individual bank with whereas the model described in this paper is trained collaboratively by different banks.

3   Methodology

3.1   Preliminaries

This section formalizes the problem setting discussed in this paper, and the FFD framework.

Definition 1 (Transaction Dataset). Let D_i denotes a credit card transaction dataset, (x_i, y_i) is the training data sample of D_i with a unique index i. Vector $x_i \in R^d$ is a d-dimensional real-valued feature vector, which is regarded as the input of the fraud detection model. Scalar $y_i \in \{0,1\}$ is a binary class label, which is the desired output of the model. $y_i = 1$ denotes that it is a fraud transaction, $y_i = 0$ denotes that it is a normal transaction.

Definition 2 (Loss Function). To facilitate the learning process, every model has a loss function defined on its parameter vector w for each data sample. The loss function captures the error of the fraud detection model on the training data. The loss of the prediction on a sample (x_i, y_i) made with the model parameters w, we define it as $\ell(x_i, y_i; w)$.

Definition 3 (Learning Rate). The learning rate controls the speed that model converges to the best accuracy. We define the learning rate as η.

Machine learning algorithm always centralizes the training data on a data center. Credit card transaction information is sensitive in both its relationship to customer privacy and its importance as a source of proprietary information for banks. Traditional machine learning models for credit card fraud detection are typically trained by individual banks with their own private dataset. Due to these datasets are privacy sensitive and large in quantity, federated learning was presented by Google [27] in 2017. Different from the traditional machine learning, federated learning enables to collaboratively learn a shared model. This shared model is trained under the coordination of a central server by using dataset distributed on the participating devices and default with privacy [28]. A typical federated algorithm – FederateAveraging (FedAvg) algorithm based on deep learning was introduced. FedAvg algorithm combines local stochastic gradient descent (SGD) on each client with a central server that performs model averaging [29]. Each client is used as nodes performing computation in order to update global shared model maintained by the central server. Every client trained their own model by using local training dataset which is never uploaded to the central server, but the update of model will be communicated. Federated learning can be concluded to five steps [27]: (1) Participating device downloads the common model from the central server. (2) Improving the model by learning data on local device. (3) Summarizes the changes of the model as a small focused update and send it using encrypted communication to the central server. (4) The server immediately aggregates with

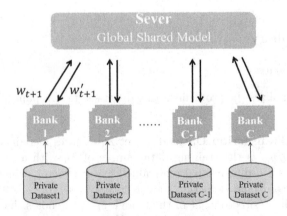

Fig. 1. Diagram of the Federated learning Framework. w_{t+1} represents the banks parameter that upload to server, w'_{t+1} represents the parameter that averaging by server.

other device updates to improve the shared model. 5) The process repeats until convergence. The structure of federated learning is illustrated in Fig. 1.

3.2   Federated Fraud Detection Framework

There are fixed set of C banks(or financial institutions) as participants, each bank possesses a fixed private dataset $D_i = \{x_i^c, y_i^c\}$ (c=1,2,3,...,C). x_i^c is the feature vector, y_i^c is the corresponding label and n_c is the size of dataset associated with participant bank c. Credit card transaction data is skewed, fraudulent transactions have a very small percentage of total number of dataset, which might cause obstructions to the performance of credit card FDS. A data level method– SMOTE [30] is selected for data rebalancing at D_i. SMOTE oversamples the minority class by generating synthetic minority examples in the neighborhood of observed ones. It is easier to implement and does not lead to increase training time or resources compared to algorithm level approach [18].

In our fraud detection system with federated learning, the goal is to allow different banks can share dataset to build an effective fraud detection model without revealing the privacy of each bank's customers. Before getting involved in training the fraud detection model, all banks will first agree on a common fraud detection model (the architecture of the model, activation function in each hidden layer, loss function, etc). For a non-convex neural network model objective is:

$$\min_{w \in \mathbb{R}^d} \quad \ell(x, y; w) \quad where \quad \ell(x, y; w) \overset{def}{=} \frac{1}{n} \sum_{i=1}^{n} \ell(x_i, y_i; w). \tag{1}$$

In federated fraud detection model, There are C banks as participant with a fixed dataset $|D_i| = n_c$, We use n to represent all the data samples involved in the whole FDS. Thus n= $\sum_{i=1}^{C} |D_i| = \sum_{c=1}^{C} n_c$. We can re-write the objective (1) as

$$\ell(x, y; w) = \quad where \quad L_c(x_c, y_c; w) = \frac{1}{n_c} \sum_{i \in D_i} \ell(x_i^c, y_i^c; w) \tag{2}$$

The server will initialize the fraud detection model parameters. At each communication round t=1,2,....., a random fraction F of banks will be selected. These banks will communicate directly with the server. First, download the current global model parameters from the server. Then, every bank computing the average gradient of the loss f_c on their own private dataset at current fraud detection model parameters w_t with a fixed learning rate η, $f_c = \nabla L_c(x_c, y_c; w)$. These banks update their fraud detection model synchronously and send the update of fraud detection model to server.

The server aggregates these updates and improves the shared model

$$w_{t+1} \leftarrow w_t - \eta \nabla \ell(x, y; w) \tag{3}$$

$$w_{t+1} \leftarrow w_t - \eta \sum_{c=1}^{C} \frac{n_c}{n} \nabla L_c(x_c, y_c; w) \tag{4}$$

$$w_{t+1} \leftarrow w_t - \eta \frac{n_c}{n} f_c \tag{5}$$

For every bank c, $w_{t+1}^c \leftarrow w_t - \eta f_c$, since (5), then

$$w_{t+1} \leftarrow w_t - \sum_{c=1}^{C} \frac{n_c}{n} w_{t+1}^c \tag{6}$$

Considering the impact of skewed data on model performance, we use the combination of data size and detection model performance α_{t+1}^c on each bank as the weight of parameter vector. it can be written as

$$w_{t+1} \leftarrow w_t - \sum_{c=1}^{C} \frac{n_c}{n} \alpha_{t+1}^c w_{t+1}^c \tag{7}$$

Increasing the weight of strong classifiers and make it plays a more important role to form a better global shared model. Each bank takes a step of gradient descent and evaluates on fraud detection model using its own credit card transactions. Then, the server applies them by taking a weighted average and makes them available to all participated banks. The whole process will go on for T iterations.

Algorithm 1. FFD framework. The C banks are index by n; B is the local minibatch size, E is the number of local epochs, and η is the learning rate.

Input: The private dataset of banks and financial institutions
Output: A credit card fraud detection model with federated learning
   ServerUpdate :
      initialize the detection classifier and its parameters w_0
      for each round t=1,2,...T do:
         Random choose max(F*C, 1) banks as N_t
         for each banks c $\in N_t$ in parallel do
         $w_{t+1}^c, \alpha_{t+1}^c \leftarrow \boldsymbol{BankUpdate}(n, w_t)$
      $w_{t+1} \leftarrow \sum_{t=1}^{T} \frac{n_c}{n} \alpha_{t+1}^c w_{t+1}^n$

   BankUpdate(n, w) :
      Data processing: rebalance raw dataset with SMOTE and split them into two part: 80% training data and 20% testing data
      Training:
         $B \leftarrow$ split D_n into baches of size B
         for each local epoch i from 1 to E do
            for batch b $\in B$ do
               w$\leftarrow$ w - $\eta \nabla \ell$(x,y;w)
      Testing
      return w and validation accuracy α to server

The increasing concern over data privacy imposes restrictions and barriers to data sharing and make it difficult to coordinate large-scale collaborative constructing a reliable FDS. Credit card FDS based on federated learning is proposed, it enables each bank to train a fraud detection model from data distributed across multiple banks. It not only helps credit card FDS learn better patterns of fraud and legitimate transactions but also protect the datasets' privacy and security. For federated optimization, communication cost is a major challenge. On the one hand, banks should fetching initial fraud detection model parameters from server. At the same time, banks should upload the update of model to server. So communication cost in FDS is symmetric. It is influenced by upload bandwidth, but in our FDS, the communication cost is related three key parameters: F, the fraction of banks that be selected to perform computation on each round; B, the minibatch size used for banks update. E, the number of local epochs. Controlling communication cost by tuning these three parameters which means we can add computation by using more banks to participate to increase parallelism or performing more computation on each bank between every communication round. The details of our fraud detection model training process are described in Algorithm 1.

4   Experimental Results

This section is organized as three parts. Firstly, we introduce the dataset that used in our FDS. Secondly, we show the performance measurement of our fraud detection model. Finally, we demonstrate the results of our experiments.

4.1   Dataset Description

We conducted a series of comprehensive experiments to manifest the superiority of our method. The experiment dataset from the European Credit Card (ECC) transactions made in September 2013 by European cardholders and it provided by the ULB ML Group [31]. This dataset contains anonymized 284,807 total transactions spanning over a period of two days, but there are only 492 fraudulent transactions in this dataset with a ratio of 1:578. The dataset is highly imbalanced as it has been observed only 0.172% of the transactions are fraudulent. Due to confidentiality issues, the original features, and some background information about the dataset cannot be provided. So this dataset contains only 30 numerical input variables which are a result of the Principal Component Analysis(PCA) transformation. It is described in Table 1. This is a classic example of an unbalanced dataset of credit card fraud(Fig. 2), it is very necessary to rebalance the raw data to prevent the classifiers from over-fitting the legitimate class and ignore the patterns of frauds.

Table 1. Credit card dataset

Normal	Fraud	Features	Instance
284315	492	30	284807

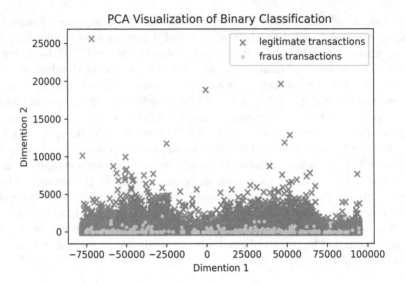

Fig. 2. Dataset visualization via PCA.

4.2 Performance Measures

Measuring the success of machine learning algorithm is a crucial task so that the best parameters suitable for credit card fraud detection system can be selected [32]. When the dataset is significantly imbalanced, accuracy is not enough to measure the performance of FDS. Accuracy will have a high value even if the FDS mispredict all instances to legitimate transactions. Therefore, we take other measures into consideration namely precision, recall, F1 and AUC which are calculated based on Table 2 where Positive correspond to fraud samples and Negative correspond to legitimate samples. Accuracy indicates the total experimental records have been classified correctly by FDS. Precision rate is a measurement of reliability of the FDS while recall rate measures the efficiency of FDS in detecting all fraudulent transactions. F1 is the harmonic mean of recall and precision. Additionally, Area Under Curve(AUC) refers to the area under the Receiver Operating Characteristic(ROC) curve, which can better describe the performance of classifiers trained with unbalanced samples.

$$Accuracy = \frac{TP + TN}{TP + FP + TN + FN} \tag{8}$$

$$Precision = \frac{TP}{TP + FP} \tag{9}$$

$$Recall = \frac{TP}{TP + FN} \tag{10}$$

$$F1 = \frac{2 \times Precision \times Recall}{Precision + Recall} \tag{11}$$

Table 2. Performance matrix

Predit	Real	
	Positive(Fraud)	Negative(Normal)
Positive(Fraud)	True Positive (TP)	False Positive (FP)
Negative(Normal)	False Negative (FN)	True Negative (TN)

4.3 Results and Discussions

In this section, a series of experiments are conducted to show the advancement of our fraud detection system. All the experiments are running on a standard server on Intel E5 with 28 CPU cores, 2.00 GHz, and 128 GB RAM. The shared global model is a CNN [33] with two convolution layers, the first with 32 channels and the second with 64 channels, each layer followed with a max pooling, a fully connected layer with 512 units and RELU activation, and a final softmax output layer.

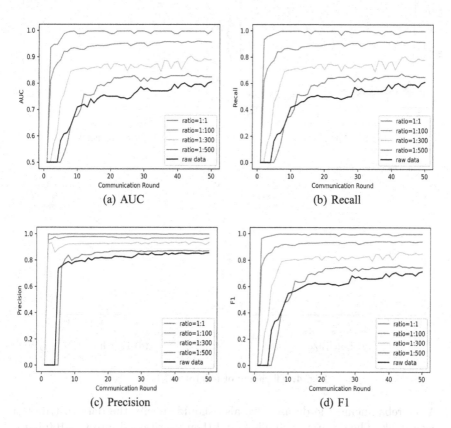

Fig. 3. Sensitive test of sampling ratio of fraud and legitimate transactions.

To minimize the impact of over-fitting, we split the dataset into 80% training data and 20% testing data. Data level approach–SMOTE is selected to rebalance raw dataset. We conduct a series of experiments on different sampling ratio with a default $E = 5$, $B = 80$ and $\eta = 0.01$. Figure 3 shows that the federated FDS with data balance mechanism outperforms FDS that trained with raw data. The better fraud detection system performed with a higher proportion of fraud transactions. Due to FDS can learn better patterns of fraud and legitimate transaction when the data is more balance. Figure 4(b) depict that when the sampling ratio is 1:1 which refers to the ratio of fraud and legitimate transactions over 1:1, the training time increased sharply but there is only a small advantage to FDS performance. Taking the training time and realistic application into consideration, we choose the sampling ratio of 1:100 to achieve an efficient FDS. From real business perspective, the average cost of misjudging 100 normal transactions is approximately the same as the mean cost of missing a fraudulent transaction.

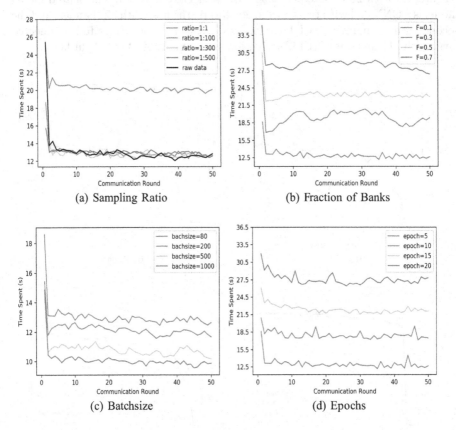

(a) Sampling Ratio

(b) Fraction of Banks

(c) Batchsize

(d) Epochs

Fig. 4. Efficiency of federated FDS.

After rebalancing the dataset, we also should specify the data distribution on each bank. The dataset is shuffled, and then partitioned into $C = 100$ banks randomly. Because the amount of transactions owned by each bank is different in reality, each bank receives a different amount of transactions. Then, the experiments with the fraction of banks F which controls the amount of banks parallelism are implemented. Table 3 demonstrates the impact of varying F for credit card fraud detection system. We calculate the number of communication round to reach a target AUC of 95.9%. The first line of Table 3 demonstrates that with the increasing Banks involved in parallel computing, the number of communication round required to reach the target AUC decreased, but the performance of FDS has become better. Time efficiency is also essential to an effective FDS which should be able to deal with limited time resource. In our FDS, the training time of every communication round (Fig. 4(a)) shows an improvement in increasing fraction of banks, but there is small advantages in performance. In order to keep a good balance between the performance of FDS and computational efficiency, in remainder experiments, we fixed $F = 0.1$.

Table 3. Sensitive test of fraction of banks

	$F = 0.1$	$F = 0.3$	$F = 0.5$	$F = 0.7$
Communication rounds	35	30	28	18
Best AUC	0.9555	0.9603	0.9638	0.9690
Best F1	0.9393	0.9441	0.9448	0.9534
Time/round(s)	12.94	19.45	23.27	28.53

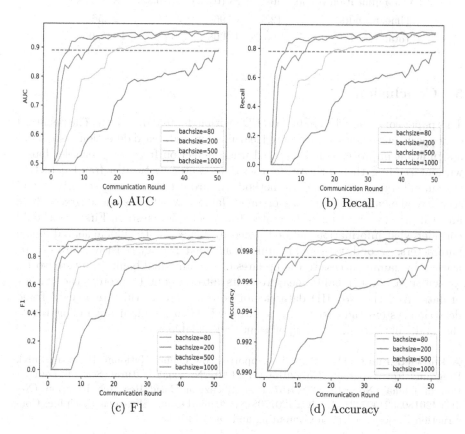

(a) AUC

(b) Recall

(c) F1

(d) Accuracy

Fig. 5. Sensitive test of local batch size.

With $F = 0.1$, adding more computation per bank on each round by decreasing batch size or increasing epochs. For batch size $-B$, we calculate the number of communication rounds necessary to achieve a target recall of 78%, F1 of 87%, AUC of 89% and validation accuracy of 99%. The results are depicted in Fig. 5, where the grey lines stand for the targets. In Fig. 4(c), smaller batch size lead longer training time on average. But the number of communication rounds to reach the targets is decreased with the increasing computation per bank by decreasing the local batch size of banks. So the total time cost is still decreased. Smaller batch size speeds the convergence and improves the performance of FDS. For local epochs, Fig. 4(d) shows that larger epoch leads the increment of training time to per communication round. But Table 4 depicts that the number to reach the target AUC of 96% is decreased. Figure 5 and Table 4 reveal that add more local SGD updates by decreasing batch size or increasing epochs per round to each bank result in a speed up to convergence rate and less computation cost.

Table 4. Sensitive test of number of local epochs

	$E = 5$	$E = 10$	$E = 15$	$E = 20$
Communication rounds	46	23 (0.5×)	15 (0.33×)	8 (0.17×)
Time/round(s)	12.98	17.96	22.38	27.56
Total Time(s)	596.08	413.08	335.7	220.48

5   Conclusion

This paper constructed a credit card FDS with federated detection. The results of our experiments show that federated learning for credit card detection system has a significant improvement. Federated fraud detection framework enables banks without sending their private data to data center to train a fraud detection system. This decentralized data method can protect the dataset sensitivity and security, alleviate the influence of unavailable dataset to some degree. There are still privacy problems in federated fraud detection system. First, we should consider what information can be learned by inspecting the global shared model parameters. Second, we should think about what privacy-sensitive data can be learned by gaining access to the updates of an individual bank. In future works, we will take more reliable measurements into account to protect the privacy of data. And the Non-IID dataset can be evaluated in this credit card fraud detection system and ensure the credit card FDS to communicate and aggregate the model updates in a secure, efficient and scalable way.

Acknowledgment. This work is supported by China National Basic Research Program (973 Program, No. 2015CB352400), National Natural Science Foundation of China (No. 61572488,61572487), Shenzhen Basic Research Program (No. JCYJ201803021457 31531, JCYJ20170818163026031), and Shenzhen Discipline Construction Project for Urban Computing and Data Intelligence.

References

1. Bahnsen, A.C., Aouada, D., Stojanovic, A., Ottersten, B.: Feature engineering strategies for credit card fraud detection. Expert Syst. Appl. **51**, 134–142 (2016)
2. Zareapoor, M., Shamsolmoali, P., et al.: Application of credit card fraud detection: based on bagging ensemble classifier. Procedia Comput. Sci. **48**(2015), 679–685 (2015)
3. Bolton, R.J., Hand, D.J.: Statistical fraud detection: a review. Stat. Sci., 235–249 (2002)
4. Sahin, Y., Bulkan, S., Duman, E.: A cost-sensitive decision tree approach for fraud detection. Expert Syst. Appl. **40**(15), 5916–5923 (2013)
5. Laleh, N., Abdollahi Azgomi, M.: A taxonomy of frauds and fraud detection techniques. In: Prasad, S.K., Routray, S., Khurana, R., Sahni, S. (eds.) ICISTM 2009. CCIS, vol. 31, pp. 256–267. Springer, Heidelberg (2009). https://doi.org/10.1007/978-3-642-00405-6_28
6. Delamaire, L., Abdou, H., Pointon, J., et al.: Credit card fraud and detection techniques: a review. Banks Bank Syst. **4**(2), 57–68 (2009)
7. Abdallah, A., Maarof, M.A., Zainal, A.: Fraud detection system: a survey. J. Netw. Comput. Appl. **68**, 90–113 (2016)
8. Sherly, K., Nedunchezhian, R.: Boat adaptive credit card fraud detection system. In: 2010 IEEE International Conference on Computational Intelligence and Computing Research (ICCIC), pp. 1–7. IEEE (2010)
9. Jha, S., Guillen, M., Westland, J.C.: Employing transaction aggregation strategy to detect credit card fraud. Expert systems with applications, 39(16), 12650–12657 (2012)
10. Bahnsen, A.C., Stojanovic, A., Aouada, D., Ottersten, B.: Improving credit card fraud detection with calibrated probabilities. In: Proceedings of the 2014 SIAM International Conference on Data Mining, pp. 677–685. SIAM (2014)
11. Liu, X.-Y., Wu, J., Zhou, Z.-H.: Exploratory undersampling for class-imbalance learning. IEEE Trans. Syst. Man Cybern. Part B (Cybern.) **39**(2), 539–550 (2009)
12. Minegishi, T., Niimi, A.: Proposal of credit card fraudulent use detection by online-type decision tree construction and verification of generality. Int. J. Inf. Secur. Res. (IJISR) **1**(4), 229–235 (2011)
13. Mohammed, R.A., Wong, K.-W., Shiratuddin, M.F., Wang, X.: Scalable machine learning techniques for highly imbalanced credit card fraud detection: a comparative study. In: Geng, X., Kang, B.-H. (eds.) PRICAI 2018. LNCS (LNAI), vol. 11013, pp. 237–246. Springer, Cham (2018). https://doi.org/10.1007/978-3-319-97310-4_27
14. Van Vlasselaer, V., et al.: APATE: a novel approach for automated credit card transaction fraud detection using network-based extensions. Decis. Support Syst. **75**, 38–48 (2015)
15. Soemers, D.J., Brys, T., Driessens, K., Winands, M.H., Nowé, A.: Adapting to concept drift in credit card transaction data streams using contextual bandits and decision trees. In: AAAI (2018)
16. Žliobaitė, I.: Learning under concept drift: an overview. arXiv preprint arXiv:1010.4784 (2010)
17. Chen, R.-C., Chen, T.-S., Lin, C.-C.: A new binary support vector system for increasing detection rate of credit card fraud. Int. J. Pattern Recogn. Artif. Intell. **20**(02), 227–239 (2006)

18. Dal Pozzolo, A., Caelen, O., Le Borgne, Y.-A., Waterschoot, S., Bontempi, G.: Learned lessons in credit card fraud detection from a practitioner perspective. Expert Syst. Appl. **41**(10), 4915–4928 (2014)

19. Bian, Y., et al.: Financial fraud detection: a new ensemble learning approach for imbalanced data. In: PACIS, p. 315 (2016)

20. Bahnsen, A.C., Stojanovic, A., Aouada, D., Ottersten, B.: Cost sensitive credit card fraud detection using bayes minimum risk. In: Proceedings-2013 12th International Conference on Machine Learning and Applications, ICMLA 2013, vol. 1, pp. 333–338. IEEE Computer Society (2013)

21. Patidar, R., Sharma, L., et al.: Credit card fraud detection using neural network. Int. J. Soft Comput. Eng. (IJSCE) **1**, 32–38 (2011)

22. Syeda, M., Zhang, Y.-Q., Pan, Y.: Parallel granular neural networks for fast credit card fraud detection. In: Proceedings of the 2002 IEEE International Conference on Fuzzy Systems, FUZZ-IEEE 2002, vol. 1, pp. 572–577. IEEE (2002)

23. Lu, Q., Ju, C.: Research on credit card fraud detection model based on class weighted support vector machine. J. Convergence Inf. Technol. **6**(1) (2011)

24. Wu, C.-H., Tzeng, G.-H., Goo, Y.-J., Fang, W.-C.: A real-valued genetic algorithm to optimize the parameters of support vector machine for predicting bankruptcy. Expert Syst. Appl. **32**(2), 397–408 (2007)

25. Bolton, R.J., Hand, D.J., et al.: Unsupervised profiling methods for fraud detection. Credit Scoring and Credit Control VII, pp. 235–255 (2001)

26. Srivastava, A., Kundu, A., Sural, S., Majumdar, A.: Credit card fraud detection using Hidden Markov Model. IEEE Trans. Dependable Secure Comput. **5**(1), 37–48 (2008)

27. McMahan, B., Ramage, D.: Federated learning: Collaborative machine learning without centralized training data. Google Research Blog (2017)

28. Konečnỳ, J., McMahan, H.B., Yu, F.X., Richtárik, P., Suresh, A.T., Bacon, D.: Federated learning: Strategies for improving communication efficiency, arXiv preprint arXiv:1610.05492 (2016)

29. McMahan, H.B., Moore, E., Ramage, D., Hampson, S., et al.: Communication-efficient learning of deep networks from decentralized data, arXiv preprint arXiv:1602.05629 (2016)

30. Chawla, N.V., Bowyer, K.W., Hall, L.O., Kegelmeyer, W.P.: SMOTE: synthetic minority over-sampling technique. J. Artif. Intell. Res. **16**, 321–357 (2002)

31. ccfraud dataset. https://www.kaggle.com/mlg-ulb/creditcardfraud

32. West, J., Bhattacharya, M.: Some experimental issues in financial fraud mining. In: ICCS 2016, pp. 1734–1744 (2016)

33. LeCun, Y., Bottou, L., Bengio, Y., Haffner, P.: Gradient-based learning applied to document recognition. Proc. IEEE **86**(11), 2278–2324 (1998)

A Relation Extraction Method Based on Entity Type Embedding and Recurrent Piecewise Residual Networks

Yuming Wang[1,3](✉), Huiqiang Zhao[1], Lai Tu[1], Jingpei Dan[2], and Ling Liu[3]

[1] School of Electronic Information and Communications,
Huazhong University of Science and Technology, Wuhan 430074, Hubei, China
`ymwang@mail.hust.edu.cn`
[2] College of Computer Science, Chongqing University, Chongqing 400044, China
[3] School of Computer Science, College of Computing,
Georgia Institute of Technology, Atlanta, GA 30332, USA

Abstract. Relation extraction is an important while challenging task in information extraction. We find that existing solutions can hardly extract correct relation when the sentence is long and complex or the firsthand trigger word does not show. Inspired by the idea of fusing more and deeper information, we present a new relation extraction method that involves the types of entities in the joint embedding, namely, Entity Type Embedding (ETE). An architecture of Recurrent Piecewise Residual Networks (RPRN) is also proposed to cooperate with the joint embedding so that the relation extractor acquires the latent representation underlying the context of a sentence. We validate our method by experiments on public data set of New York Times. Experiment results show that our method outperforms the state-of-the-art models.

Keywords: Relation extraction · Entity Type Embedding · Recurrent neural networks · Residual Networks

1 Introduction

Information extraction aims at extracting structured information from large-scale text corpora, of which the main task is to recognize entities, the relation of the entity pair and the events involved [1]. The extracted facts of the relations can support a wide range of applications like Semantic Search, QA and etc [2]. To extract relations, most of the current solutions employ supervised training methods to learn from labelled corpus. These solutions usually use word embedding [3,4] and position embedding [5] in the representation layer. The embedded vectors then pass through a neural network and they are jointly trained with the labelled data. However, based on our experiment on the New York Times corpus data, existing attention-based solutions fail to extract relation in the following challenging cases as illustrated in Fig. 1.

© Springer Nature Switzerland AG 2019
K. Chen et al. (Eds.): BigData 2019, LNCS 11514, pp. 33–48, 2019.
https://doi.org/10.1007/978-3-030-23551-2_3

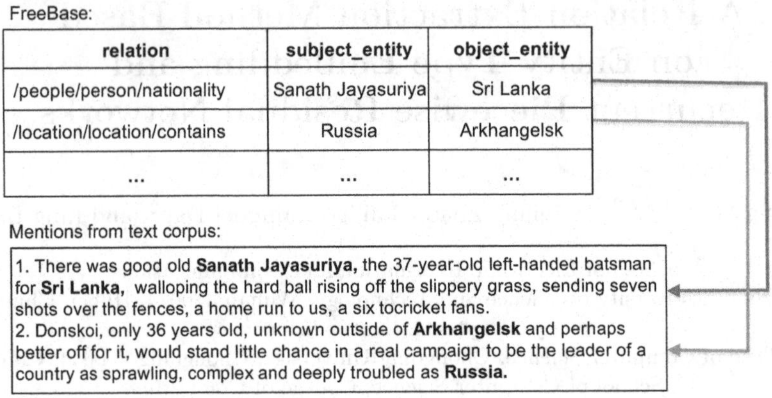

FreeBase:

relation	subject_entity	object_entity
/people/person/nationality	Sanath Jayasuriya	Sri Lanka
/location/location/contains	Russia	Arkhangelsk
...	...	...

Mentions from text corpus:

1. There was good old **Sanath Jayasuriya,** the 37-year-old left-handed batsman for **Sri Lanka,** walloping the hard ball rising off the slippery grass, sending seven shots over the fences, a home run to us, a six tocricket fans.
2. Donskoi, only 36 years old, unknown outside of **Arkhangelsk** and perhaps better off for it, would stand little chance in a real campaign to be the leader of a country as sprawling, complex and deeply troubled as **Russia.**

Fig. 1. An example of relation extraction model.

1. Cases that sentences have no firsthand trigger word [6] of the corresponding relation, such as the first example in Fig. 1.
2. Cases that the sentences are long and the position of the subject entity is far apart from the object entity, such as the second example in Fig. 1.

We observe that the extraction failure is due to the missing of latent information hidden deep in the context. To address this issue, we propose a method based on joint embedding and modified residual network. First we involve extra knowledge in the representation layer. More specifically, we introduce an Entity Type Embedding (ETE) module that maps the types of subject and object entities into vectors. The vectors then pass through a CNN network and are added to the encoded vector of word and position embedding with trained weights. Secondly, we design an new encoder module based on Recurrent Piecewise Residual Networks (RPRN). The architecture of RPRN increases the depth of the neural network so that the encoder is able to extract latent information from the sentence.

We evaluate our method on the New York Times dataset both qualitatively and quantitatively. Results show that our method can correctly extract the relations in the challenging cases that existing attention-based solutions fail. And the proposed solution also provides a better overall performance over all baseline methods.

In the following sections, we begin with a brief review of the related work in Sect. 2. And then, in Sect. 3, we represent our approach for relation extraction and elaborate the core procedure of our algorithm. Section 4 provides quantitative and qualitative experimental results. Finally, we conclude our work in Sect. 5.

2   Related Work

For relation extraction, current supervised method may extract more features and obtain higher precision and recall. However, the main problem of such method is time consuming and requiring intensive labor of labeling corpus manually. In response to the limitation above, Mintz et al. [7] applied Distant Supervision (DS) to relation extraction, they aligned the New York Times (NYT) corpus with the large-scale knowledge base of Freebase. DS assumes that the entity pair mentioned in a sentence implies the semantic relation of it in knowledge base. Riedel et al. [8] relaxed the assumption of DS, of which error rate reduced by 31% compared to Mintz. To break through the limitation that DS assumes each entity pair corresponds to only one relation, Hoffmann et al. [9] proposed a novel model of Multi-Instance Multi-Label to proceed relation extraction, which allows for the scenario existing multiple relations between an entity pair. As for the serious flaw of generating negative examples, Bonan et al. [10] proposed an algorithm that learns from only positive and unlabeled labels at the pair-of-entity level.

Although having achieved preferable results, the methods based on traditional machine learning rely on preprocessing such as Part-of-Speech tagging (POS) [11], Semantic Role Labeling (SRL) [12] and so on. Errors may exist during the preprocessing and may be propagated and amplified in relation extraction.

It is of much concern that deep learning allows computational models that are composed of multiple processing layers to learn representations of data with multiple levels of abstraction [13]. Zeng et al. [14] first adopted Convolutional Neural Networks (CNN) while modeling relation extraction. To tackle the time consuming and manual labeling issues, the authors further integrated a new CNN with DS for relation extraction in [15].

Miwa et al. [16] proposed an end-to-end neural network, which utilized bidirectional Long Short Term Memory (LSTM) [17] network to model relation extraction. Peng et al. [18] and Song et al. [19] explored a general relation extraction framework based on graph LSTM that can be easily extended to cross-sentence n-ary relation extraction. To alleviate performance degradation caused by noise instance, Lin et al. [20] applied selective attention mechanism to relation extraction, which pledges that the weight of effective instance will rise and the noise one will decline. In addition, Wang et al. [21] and Du et al. [22] proposed a novel multi-level attention mechanism for relation extraction. Wu et al. [23] and Qin et al. [24] introduced the process of adversarial training to relation extraction to further enhance its generalization capacity.

3   Methodology

In this section, we represent the details of our approach, and elaborate the core procedure of our algorithm in the form of pseudo code.

3.1 Overview

The overall architecture of our model is sketched in Fig. 2, which is mainly divided into the following parts:

1. Embedding layer: Word embedding, position embedding and Entity Type Embedding (ETE) work jointly as the distributed representation of the model;
2. Encoder Layer: Word and position embedding output is encoded by a RPRN based encoder and ETE output is encoded with a CNN based encoder;
3. Attention Layer: To relieve the performance degradation of noisy instances, we employ a selective attention mechanism to pick out the positive instances while training.

Take the instance in Fig. 2 as an example, the top table represents the word embedding and position embedding, where word embedding is pre-trained and position embedding is initialized randomly, and the bottom table represents the ETE. The embedding of concatenating word with position goes through the architecture of RPRN, and then concatenates with the output of encoding ETE. After encoding, the instances possessing the same entity pair will be put into the same bag, and then the attention based selector is going to pick out positive instances from the bag as far as possible for the training relation. Finally, we will get a vector representing the latent pattern of the training relation, on which the last layer of classifier bases to output the most probable relation.

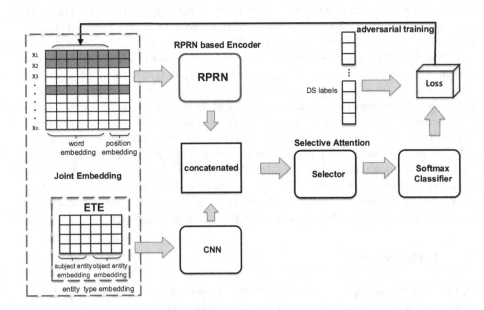

Fig. 2. The overall architecture of our model for relation extraction.

3.2   Joint Embedding

The distributed representation layer of relation extraction composes of word embedding, position embedding and ETE, whether it represents adequately makes a difference directly on the effects of the model.

(1) **Word Embedding.** Word embedding is a distributed representation of words to be mapped from a high-dimensional word space to a low-dimensional vector space, where the semantically similar words are close to each other as well. It can be seen that word embedding can express the semantic relation among the words. In our model, we utilize skip-gram [4] to pre-train word vector, but the pre-trained word vector is merely served as the initialization of the word embedding, afterwards it will be updated while training.

(2) **Position Embedding.** In order to exploit the information of position of each word in the sentence, position embedding [5] is applied to relation extraction. The information of position aforementioned denotes the relative distances (illustrated in Fig. 3) between each word in the sentence and the corresponding entity pair, and then map the relative distances of each word to a vector space. Besides, each word in a sentence correspond to two vectors initialized randomly, then the two vectors will be updated constantly along with continually training.

Fig. 3. The relative position between word and the corresponding entity.

(3) **Entity Type Embedding.** As can be seen from Fig. 1, the information of entity type plays an important role in relation extraction. Yet the structure of entity type is widely divergent from word embedding and position embedding, therefore it can not be served as the distributed representation layer that simply concatenating ETE with word embedding and position embedding. The scheme we adopted is that the ETE to be served as a part of representation is encoded separately with CNN, of which process is sketched in Fig. 2.

In our model, the dimension of word embedding is $d_w = 50$, the position embedding's is $d_p = 5$, and then joint together to transform an instance of training into a matrix $S \in \mathbb{R}^{n \times d}$, n = 120 denotes maximum length of sentences in the corpus, $d = d_w + 2 \times d_p$ denotes the dimension of representation layer after jointing word embedding and position embedding. The matrix S will be fed to the encoder of RPRN. The dimension of the representation for ETE is $d_{et} = 12$, the ETE represent as $E_{et} \in \mathbb{R}^{2 \times n_{et} \times d_{et}}$, where $n_{et} = 100$ denotes the maximum number of types for an entity over all corpus.

3.3   RPRN Based Encoder

Since ETE is introduced to the embedding layer and is encoded separately, the encoder in this model composes of two parts, a RPRN based encoder for

word and position embedding, and a CNN based encoder for ETE. Besides, the
architecture of RPRN is illustrated in Fig. 4.

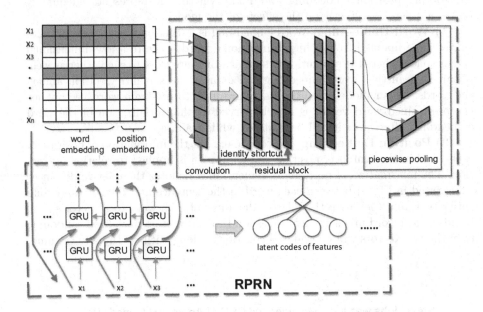

Fig. 4. The principle of the architecture of RPRN.

(1) Sequence Learning. We employ bidirectional RNN to learn the context
information of the sentence, and we use Gated Recurrent Unit (GRU) [25] as
the recurrent unit. GRU can deal with the problem of long-term dependency,
and is more streamlined than LSTM. In our model, the gated unit is simplified
into two gates, namely update gate z_t and reset gate r_t, which are derived as
Eqs. (1) and (2) respectively.

$$z_t = \sigma(W_z x_t + U_z h_{t-1} + b_z) \tag{1}$$

$$r_t = \sigma(W_r x_t + U_r h_{t-1} + b_r) \tag{2}$$

The output status of GRU is h_t, derived as Eq. (4), which is determined by
the updated and reserved information based on the last time's:

$$h_t' = g(W_h x_t + U_h(r_t \odot h_{t-1}) + b_h) \tag{3}$$

$$h_t = (1 - z_t) \odot h_{t-1} + z_t \odot h_t' \tag{4}$$

(2) Residual Networks. To learn finer features, deeper neural network is
preferred. However, deeper neural network has to face the challenge of gradient
vanishing or exploding [26], which makes the model hard to optimize. Therefore,
we use a residual network [27] based model in the encoder as illustrated in Fig. 2.

We modify the identity mapping part and pooling part where the size of identity block is $I_s = 2$, and the number of blocks is $I_n = 3$. The RPRN with CNN based encoder for ETE altogether comprise of 12 layers in this model. The residual function is represented as Eq. (5), where x is the input and H is the output function of identity block.

$$F(x) = H(x) - x \tag{5}$$

(3) Piecewise Pooling. For extracting the semantic relation between the entity pair, we employ piecewise pooling to learn structured information, which is an effective approach of structured learning [28]. We introduce mask information represented as $M \in \mathbb{R}^{m \times n}$ and its embedding represented as $M_e = [[0, 0, 0], [1, 0, 0], [0, 1, 0], [0, 0, 1]]$ to implement the piecewise pooling, where m denotes the batch size. The result after the operating of piecewise pooling is $P \in \mathbb{R}^{m \times (3*h)}$ derived as Eq. (6), where h denotes the hidden size.

$$P = \phi(\pi(\epsilon(\gamma(M_e, M) \times c, 2) + \epsilon(C, 3)) - c, [-1, 3*h]) \tag{6}$$

Regarding the parameters involved in the equation above, ϕ denotes the function to adjust the shape of tensor, π is to proceed piecewise max pooling, ϵ is to expand a certain dimension of the tensor, γ denotes the function to look up embedding of mask, c is a constant, and C is a output tensor of the convolutional part. The principle of piecewise pooling is illustrated in Fig. 5.

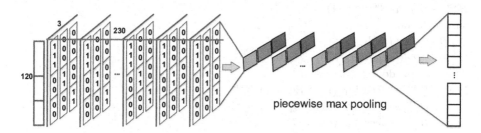

Fig. 5. The principle of piecewise pooling.

As a part of RPRN, the layer of piecewise pooling is designed to obtain principal features from the output of previous layer. Piecewise pooling, in the context of this paper, means dividing three parts according to the positions of entity pair in a sentence, and then each part proceed max pooling respectively.

The max pooling is utilized for ETE, and its hidden size is $h_{et} = 80$. The result of the operation for max pooling is then concatenated with the piecewise pooling's to be served as the encoder of relation extractor, and its output will be fed to the instance selector.

3.4 Selective Attention

The method of DS is usually applied to relation extraction to solve the problem of time consuming and labor intensive for labeling corpus manually, yet it simultaneously confronts with error labeling. The initial assumption of DS serves all the instances in a bag as positive examples, accordingly there are plenty of noise instances involving in training. For this issue, Riedel et al. [8] adopted serving the most likely one as positive example for training, which tremendously alleviates the performance reduction caused by noise instances. Meanwhile, such method discard so many effective instances and then hinder improving the model of relation extraction. Thus Lin et al. [20] applied the selective attention based mechanisms to relation extraction, and its principle is sketched in Fig. 2, the output vector A is derived as Eq. (7).

$$A = \omega\left(\mu\left\{\epsilon\left(\frac{exp^{H \cdot \gamma(R,y) + bias}}{\sum_{k=1}^{b} exp^{H_k \cdot \gamma(R_k, y_k) + bias_k}}, 0\right), H\right\}\right) \tag{7}$$

Algorithm 1. The core procedure of our algorithm

Input: training corpus $B = \{B_1, B_2, \cdots, B_m\}$, hyper-parameters
    K, pre-trained word vector V^w.
initialize parameters Θ, V^p, and V^{et}
$B^w, B^p, B^{et} \leftarrow \gamma(V^w, B^w), \gamma(V^p, B^p), \gamma(V^{et}, B^{et})$
$B^{wp} \leftarrow concat(B^w, B^p, -1)$
for l to K^l do
    $B \leftarrow shuffle(B)$
    $T \leftarrow ceil(length(B)/K^t)$
    for t to T do
        $H_{t-1}^{wp} \leftarrow$ feed Eq. (1) to Eq. 4 on B_t^{wp}
        $H_{t-2}^{wp} \leftarrow$ feed Eq. (5) on B_t^{wp}
        $H_{t-2}^{wp} \leftarrow$ feed Eq. (6) on H_{t-2}^{wp}
        $H_t^{wp} \leftarrow concat(H_{t-1}^{wp}, H_{t-2}^{wp}, -1)$
        $H_t^{et} \leftarrow cnn(B_t^{et})$
        $H_t \leftarrow concat(H_t^{wp}, H_t^{et}, -1)$
        foreach $H_{t-i} \in H_t$ do
            pick out positive instance as far as possible
            apply Eq. (7) to H_{t-i}
        end
        $O_t \leftarrow softmax(H_t)$
        then update the parameters Θ
        $\Theta \leftarrow \Theta + \nabla L\Theta(O_t)$
    end
end
Output: extracted relation B^r

In the Eq. (7) above, ω is a squeeze function to strip all the dimensions valued 1 for the tensor, μ denotes matrix multiplication, H denotes the output of encoder, R denotes relation matrix, y is the ground-truth of corpus, *bias* denotes a bias vector, b denotes the bag size, and other notations have been explained above.

Adopting the attention based mechanisms described above, the sentences that indeed express their relations involved in knowledge base can be picked out as positive instances, and those that don't will be served as noises. More specifically, such selective attention based mechanisms pledges that the weights of effective instances will rise and the noise one will decline, which effectively alleviates the performance degradation caused by noise instances.

For the last layer, we adopt softmax to classify the relation and serve cross-entropy as the loss function. Besides, we introduce the adversarial process [23] to perturb the input representation while training so that the model acquires better generalization capacity on the testing set.

In summary, the core procedure for implementing our model is elaborated in Algorithm 1. As for the notations in Algorithm 1, we will make some notes necessarily in the next moment. B, as the preprocessed training corpus, mainly contains the information of word, position and entity type, that's B^w, B^p, B^{et}. Θ denotes the parameters overall the training process of the model. V^p denotes position vector, V^{et} denotes entity type vector, which are all initialized randomly and subsequently optimized progressively with continuing training. the hyper-parameters K directly utilized in Algorithm. 1 are max epoch K^l and batch size K^t. Besides, H denotes the output of hidden layer, O denotes the final output of last layer, and L_Θ is the loss function of the entire model.

4 Experiments

In this section, we present the comparative results of our relation extraction method and several baseline solutions. We also compare the performance of different combinations of proposed sub-modules.

4.1 Experimental Setup

We conduct an experiment on the public dataset of New York Times (NYT) in this work, which is extensively used in relation extraction. The training set of NYT contains 522,611 sentences, 281,270 entity pairs, and 18,252 relation facts; The testing set contains 172,448 sentences, 96,678 entity pairs, and 1,950 relation facts. The entity pair, represented as <subject_guid#object_guid>, will obtain the score corresponding to each relation, and then it will subordinate to the relation obtaining the highest score.

(1) **Hyper-parameters Settings.** The best hyper-parameters, set as Table 1, are specified through cross validation and grid search in this work.

(2) **Measurement Metrics.** According to the combination of the prediction results and ground truth, the samples of corpus can be divided into four parts:

Table 1. Hyper-parameters settings.

Param name	Value	Description
word_dim	50	The dimension of word embedding
pos_dim	5	The dimension of pos embedding
et_dim	12	The dimension of ETE
hidden_size	230	The size of hidden layer for encoding word and position embedding
et_hidden_size	80	The size of hidden layer for encoding the information of ETE
ib_num	3	Num of identity block for RPRN
learning_rate	0.5	The learning rate for optimizer
drop_out	0.5	The drop out rate of discarding some neurons while training

True Positive (TP), False Positive (FP), False Negative (FN), True Negative (TN). We use precision, recall, F1 and AUC as the metrics which are widely used in relation extraction evaluation.

4.2 Baselines

We adopt four state-of-the-art solutions as baselines:

- **pcnn_one** (Zeng et al. [15]) served the most likely one in a bag as positive instance for training.
- **pcnn_att** (Lin et al. [20]) applied the selective attention based mechanisms to relation extraction, which effectively utilized plenty of positive instances in a bag.
- **bilstm_att** (Miwa et al. [16]) presented a novel end-to-end neural model by stacking bidirectional tree structured LSTM for relation extraction.
- **pcnn_att_ad** (Wu et al. [23], Qin et al. [24]) introduced adversarial training for relation extraction to further enhance its generalization capacity.

We also compare the performance of different combinations of proposed sub-modules, listed as follows:

- **ete_pcnn_att** utilizes ETE for the whole training data, compared to the previous model, still more proving the effectiveness of ETE.
- **ete_pcnn_att_ad** utilizes ETE and adversarial training jointly.
- **rprn_att** adopts the architecture of dubbed RPRN we devised.
- **rprn_att_ad** coalesces the architecture of RPRN and adversarial training.
- **ete_rprn_att** integrates the ETE with the architecture of RPRN we devised.
- **ete_rprn_att_ad** is a joint model based on ETE, the architecture of RPRN and adversarial training, which achieves best results in this paper.

4.3   Evaluation Results

Table 2 gives an overview of the performance of various models. Figures 6, 7, 8 and 9 further show the precision to recall performance of different method. Results show that all the metrics have been improved significantly. More specifically, the comparisons among various models reveal the following observations:

1. Fig. 6 shows that the performance of four baseline methods do not have distinct difference.
2. In Fig. 7a and b, we compare the models using ETE in two baselines respectively. We can see that ETE improves the baseline methods.
3. In Fig. 8a and b, RPRN based encoder is used alone. We find that single RPRN based encoder also improves the performance.

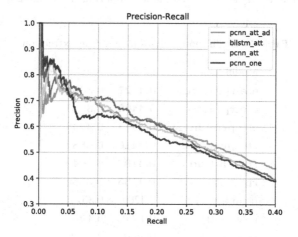

Fig. 6. Comparisons among the models of four **baselines**.

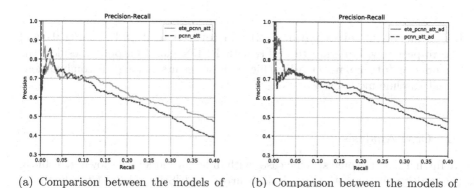

(a) Comparison between the models of **ete_pcnn_att** and **pcnn_att**.

(b) Comparison between the models of **ete_pcnn_att_ad** and **pcnn_att_ad**.

Fig. 7. The curve of PR for proposed models utilizing the ETE.

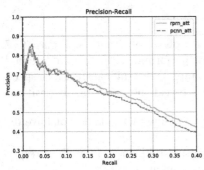
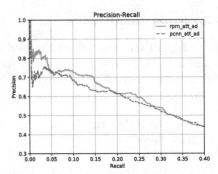

(a) Comparison between the models of **rprn_att** and **pcnn_att**.

(b) Comparison between the models of **rprn_att_ad** and **pcnn_att_ad**.

Fig. 8. The curve of PR for proposed models adopting the architecture of RPRN.

Table 2. Evaluation metrics for relation extraction.

Model name	F1-score	AUC-value
ete_rprn_att_ad	**0.4427**	**0.3965**
ete_rprn_att	**0.4409**	**0.3935**
ete_pcnn_att_ad	**0.4443**	**0.3898**
ete_pcnn_att	**0.4416**	**0.3882**
rprn_att_ad	**0.4198**	**0.3649**
rprn_att	**0.4135**	**0.3568**
pcnn_att_ad	0.4209	0.3518
bilstm_att	0.4020	0.3467
pcnn_att	0.3991	0.3414
pcnn_one	0.3984	0.3280

4. The models of (**ete_rprn_att_ad**) and (**ete_rprn_att_ad**), in Fig. 9a and b respectively, integrate the ETE with RPRN based encoder. Besides, the model of (**ete_rprn_att_ad**) achieves best results among all combinations and baseline methods.

4.4 Case Analysis

We further look at some specific cases with different models as shown in Table 3. The entity pair, namely 'Sanath Jayasuriya' and 'Sri Lanka', of the first sentence in Table 3 indeed express the 'nationality' relation, which is comparatively obscure and it is hard to be extracted purely through the information of the context in the sentence. The model (**pcnn_att**), quite understandably, can not

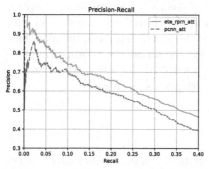

(a) Comparison between the models of ete_rprn_att and pcnn_att.

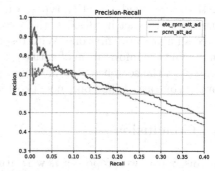

(b) Comparison between the models of ete_rprn_att_ad and pcnn_att_ad.

Fig. 9. The curve of PR for proposed models integrating ETE and RPRN.

Table 3. Cases for relation extraction.

Sentence	Extraction result	
	ete_pcnn_att	pcnn_att
There was good old **Sanath Jayasuriya**, the 37-year-old left-handed batsman for **Sri Lanka**, walloping the hard ball rising off the slippery grass, sending seven shots over the fences, a home run to us, a six to cricket fans	/people/person/nationality	NA
Donskoi, only 36 years old, unknown outside of **Arkhangelsk** and perhaps better off for it, would stand little chance in a real campaign to be the leader of a country as sprawling, complex and deeply troubled as **Russia**	/location/location/contains	NA
In 1948, Rabbi Kret came to **New York City**, and with the help of Rabbi Yosef Eliyahu Henkin, of blessed memory, he was hired as the rabbi of the Old Broadway Synagogue in the West Harlem neighborhood of **Manhattanville**	/location/neighborhood /neighborhood_of	NA
But Justice Michael R. Ambrecht of State Supreme Court in Manhattan said that as a professional BASE -LRB- Bridge, Antenna, Span, Earth -RRB- jumper, Mr. Corliss, who has parachuted from the Eiffel Tower, the Golden Gate Bridge and the Petronas Towers in **Kuala Lumpur, Malaysia**, was experienced and careful enough to jump off a building without endangering his own life or anyone else's	/location/administrative_division /country	/people/person /nationality

extract such relation, whereas ours (**ete_pcnn_att**) leveraging the information of the corresponding entity types, which are severally 'person' and 'country', manage to extract it. Besides, the second and the third sentences may be analyzed similar to the first one. With respect to the fourth sentence, the model not employing the information of entity type is puzzled about the information of the context in the sentence, and has improperly identified the 'nationality' relation on the contrary. While our model correctly extract the relation.

5 Conclusion

In this paper, we propose a new relation extraction method that involves two techniques. We first present an entity type embedding module that integrates extra type information of subject and object entities to help eliminate word-sense ambiguity and emphasize the hidden relation between the two types of entities. We then design a new architecture of encoder based on GRU and residual network. The RPRN based encoder exploits the benefit of deeper neural network to learn the context of a sentence and extract finer syntactic and semantic features.

Experiment results show that both techniques improve the performance of relation extraction separately and they provide best result when used together. In the future work, we plan to apply the architecture to Semantic Role Labeling, Named Entity Recognition and so on, and then train the model of relation extraction and these tasks jointly to improve them together.

Acknowledgement. The authors from Huazhong University of Science and Technology, Wuhan, China, are supported by the Chinese university Social sciences Data Center (CSDC) construction projects (2017–2018) from the Ministry of Education, China. The first author, Dr. Yuming Wang, is a visiting scholar at the School of Computer Science, Georgia Institute of Technology, funded by China Scholarship Council (CSC) for the visiting period of one year from December 2017 to December 2018. Prof. Ling Liu's research is partially supported by the USA National Science Foundation CISE grant 1564097 and an IBM faculty award. Any opinions, findings, and conclusions or recommendations expressed in this material are those of the author(s) and do not necessarily reflect the views of the funding agencies.

References

1. Zhang, Q., Chen, M., Liu, L.: A review on entity relation extraction. In: International Conference on Mechanical, Control and Computer Engineering, pp. 178–183 (2017)
2. Kumar, S.: A survey of deep learning methods for relation extraction. CoRR abs/1705.03645 (2017)
3. Mikolov, T., Sutskever, I., Chen, K., Corrado, G.S., Dean, J.: Distributed representations of words and phrases and their compositionality. In: Advances in Neural Information Processing Systems, pp. 111–3119 (2013)
4. Leimeister, M., Wilson, B.J.: Skip-gram word embeddings in hyperbolic space. CoRR abs/1809.01498 (2018)

5. Shi, W., Gao, S.: Relation extraction via position-enhanced convolutional neural network. In: 2017 International Conference on Intelligent Environments (IE), pp. 142–148 (2017)
6. Ding, B., Wang, Q., Wang, B.: Leveraging text and knowledge bases for triple scoring: an ensemble approach - the Bokchoy triple scorer at WSDM Cup 2017. CoRR abs/1712.08356 (2017)
7. Mintz, M., Bills, S., Snow, R., Jurafsky, D.: Distant supervision for relation extraction without labeled data. In: ACL/IJCNLP (2009)
8. Riedel, S., Yao, L., McCallum, A.: Modeling relations and their mentions without labeled text. In: Balcázar, J.L., Bonchi, F., Gionis, A., Sebag, M. (eds.) ECML PKDD 2010. LNCS (LNAI), vol. 6323, pp. 148–163. Springer, Heidelberg (2010). https://doi.org/10.1007/978-3-642-15939-8_10
9. Hoffmann, R., Zhang, C., Ling, X., Zettlemoyer, L.S., Weld, D.S.: Knowledge-based weak supervision for information extraction of overlapping relations. In: ACL (2011)
10. Min, B., Grishman, R., Wan, L., Wang, C., Gondek, D.: Distant supervision for relation extraction with an incomplete knowledge base. In: HLT-NAACL (2013)
11. Goldwater, S.: Part of speech tagging. In: Encyclopedia of Machine Learning and Data Mining (2017)
12. Tan, Z., Wang, M., Xie, J., Chen, Y., Shi, X.: Deep semantic role labeling with self-attention. In: AAAI (2018)
13. LeCun, Y., Bengio, Y., Hinton, G.E.: Deep learning. Nature **521**, 436–444 (2015)
14. Zeng, D., Liu, K., Lai, S., Zhou, G., Zhao, J.: Relation classification via convolutional deep neural network. In: COLING (2014)
15. Zeng, D., Liu, K., Chen, Y., Zhao, J.: Distant supervision for relation extraction via piecewise convolutional neural networks. In: EMNLP (2015)
16. Miwa, M., Bansal, M.: End-to-end relation extraction using LSTMs on sequences and tree structures. CoRR abs/1601.00770 (2016)
17. Yao, K., Cohn, T., Vylomova, E., Duh, K., Dyer, C.: Depth-gated LSTM. CoRR abs/1508.03790 (2015)
18. Peng, N., Poon, H., Quirk, C., Toutanova, K., Yih, W.T.: Cross-sentence N-ary relation extraction with graph LSTMs. TACL **5**, 101–115 (2017)
19. Song, L., Zhang, Y., Wang, Z., Gildea, D.: N-ary relation extraction using graph-state LSTM. In: EMNLP (2018)
20. Lin, Y., Shen, S., Liu, Z., Luan, H., Sun, M.: Neural relation extraction with selective attention over instances. In: ACL (2016)
21. Wang, L., Cao, Z., de Melo, G., Liu, Z.: Relation classification via multi-level attention CNNs. In: ACL (2016)
22. Du, J., Han, J., Way, A., Wan, D.: Multi-level structured self-attentions for distantly supervised relation extraction. In: EMNLP (2018)
23. Wu, Y., Bamman, D., Russell, S.J.: Adversarial training for relation extraction. In: EMNLP (2017)
24. Qin, P., Xu, W., Wang, W.Y.: DSGAN: generative adversarial training for distant supervision relation extraction. In: ACL (2018)
25. Dey, R., Salem, F.M.: Gate-variants of gated recurrent unit (GRU) neural networks. In: 2017 IEEE 60th International Midwest Symposium on Circuits and Systems (MWSCAS), pp. 1597–1600 (2017)
26. Pascanu, R., Mikolov, T., Bengio, Y.: On the difficulty of training recurrent neural networks. In: ICML (2013)

27. He, K., Zhang, X., Ren, S., Sun, J.: Deep residual learning for image recognition. In: 2016 IEEE Conference on Computer Vision and Pattern Recognition (CVPR), pp. 770–778 (2016)
28. Rohekar, R.Y.Y., Gurwicz, Y., Nisimov, S., Koren, G., Novik, G.: Bayesian structure learning by recursive bootstrap. CoRR abs/1809.04828 (2018)

Research on Parallel Computing Method of Hydrological Data Cube

Rong Wang[1(✉)], Dingsheng Wan[1], Zixuan Zhang[2], and Qun Zhao[1]

[1] College of Computer and Information, Hohai University,
Nanjing 211100, China
rongwang@hhu.edu.cn
[2] Bank of Jiangsu, Nanjing 211100, China
zhangzixuan222@163.com

Abstract. Hydrological data is characterized by large data volume and high dimension. In view of non-hierarchical dimension data in hydrological data, this paper proposed Segmented Parallel Dwarf Cube (SPD-Cube). Firstly, a high-dimensional cube is divided into several low-dimensional segments. Then by calculating the sub-dwarf of each low-dimensional segment, the storage space required for cube materialization is greatly reduced. Finally, combining Map/Reduce technology to construct, store and query several sub-dwarfs in parallel. Experiments show that SPD-Cube not only realizes highly compressed storage of data, but also realizes efficient query by combining Map/Reduce distributed parallel architecture.

Keywords: Hydrological data · Dwarf cube · Map/Reduce

1 Introduction

With the construction and development of informatization in water conservancy industry, China has abundant basic hydrological data. In the face of complex hydrological data, the single data storage management method can no longer meet the users' needs for subject-oriented rapid analysis and query, which is not conducive to human-computer interaction and statistical analysis. In a data warehouse, data is usually organized and stored in the form of a data cube [1, 2]. As the main method of multi-dimensional data analysis, data cube provides users with multi-angle conceptual views, which can effectively support the complex analysis operations of Multi-dimension On-Line Analytical Processing (MOLAP) [3, 4], such as winding up, drilling down, slicing and cutting. Therefore, the hydrological data is expressed in the form of data cube, so that the characteristics of hydrological data can be expressed in a more intuitive and friendly manner. The deep processing and analysis of hydrological data has important application research value for water resources development and utilization, soil erosion prevention and control, and hydrological infrastructure construction.

Xin Dong, Han Jiawei et al. proposed the Star-Cubing method based on the iceberg cube [5], this method uses the data structure of the star tree to store the aggregation unit, and combines the multi-channel array aggregation method [6, 7] and the pruning strategy [8] in the traditional BUC algorithm [9], which effectively improves the

K. Chen et al. (Eds.): BigData 2019, LNCS 11514, pp. 49–64, 2019.
https://doi.org/10.1007/978-3-030-23551-2_4

efficiency of data retrieval, but at the same time it is affected by the order of dimensions; Based on the advantages of BUC and Star-Cubing method [10], Shao Zheng proposed the MM-Cubing method [11], which adopts different calculation methods according to the number of occurrences of dimension attributes in the data table; In order to compress the space of the cube, Li Shengen, Wang Shan et al. proposed the concept of a closed cube [12], which divides the elements in the cube into closed units and non-closed units. Subsequently, Xiao Weiji et al. proposed an effective closed cube indexing technique—the CC-Bitmaps method [13], which uses the ideas of prefix sharing, suffix sharing and bit encoding to further compress the data storage space; In order to solve the problem of losing basic semantic relations when compressing data cubes, Laks v. s. Lakshmanan et al. proposed the concept of a quotient cube [14], which divides the units aggregated from the same unit into an equivalent class, and these units have the same aggregate value; Wang Wei et al. proposed the concept of Condensed cube [15], for cells aggregated by the same tuple, the condensed cube removes these redundant cells with the same aggregate value, leaving only one basic unit. At present, there have been a lot of researches on cube materialization technology at home and abroad. These results can effectively reduce the storage space and query response time of cubes. But after research, it is found that for data with multi-dimensional and hierarchical characteristics such as hydrological data, cube calculation efficiency is still low and cannot meet the demand of efficient query analysis. In order to solve the problem of partial information redundancy in the condensed cube, Yannis Sismanis et al. proposed the concept of Dwarf cube through research. This method adopts a special directed acyclic graph structure to reduce prefix redundancy and suffix redundancy. Dwarf cube can improve computing efficiency to a certain extent, but it still cannot meet the requirements of efficient query analysis. Therefore, some efficient distributed computing mechanisms should be combined to meet the needs of users.

For non-hierarchical data with many dimensions, the traditional cube computing method has a certain degree of redundancy, and with the increase of dimensions, the number of aggregation units will increase exponentially. Therefore, Segmented Parallel Dwarf Cube (SPD-Cube) is proposed to guarantee the data's MOLAP efficiency, balanced storage space and query efficiency. In view of the characteristics of non-hierarchical and high-dimensional hydrological data, this paper studies the calculation method, query method and incremental update method on basis of improved dwarf cube.

2 SPD-Cube

2.1 Dwarf Cube

According to the characteristics of hydrological data and the segmentation idea of data cube, the multidimensional hydrological data are divided into segments according to dimensions. The non-hierarchical dimension of the daily rainfall fact table in the hydrological data is given, and three commonly used non-hierarchical dimension attributes are selected into one segment: grade of hydrometric station (STGRD), grade of flood-reporting station (FRGRD) and item code of observation (OBITMCD), precipitation (P) is the measurement value, TID represents the table ID, the TID-List is an inverted index set corresponding to the tuple, as shown in Table 1.

Table 1. Rainfall fact table of non-hierarchical dimension fragments

TID	STGRD	FRGRD	OBITMCD	...	P (mm)
1	S1	F1	P	...	0.5
2	S1	F2	PE		0.2
3	S2	F3	PE	...	2.3
4	S2	F3	P	...	0.6
5	S3	F1	P	...	1.9
...	...	...	...	...	...

Definition 2.1 Dwarf Cube. The dwarf cube uses a directed acyclic graph structure to store data. Prefix Expansion and Suffix Coalescing are used to reduce the redundant prefixes and suffixes in the data cube so as to solve the problem of explosive growth of cube storage space.

A complete dwarf cube needs to meet the following characteristics:

(1) A dwarf cube consisting of a basic table of N-dimensions has N layers in the directed acyclic graph, the graph has only one root node;
(2) In addition to leaf nodes, each node contains several binary groups, which exist in the form of [Key: Point], the keyword Key represents the attribute value of the non-hierarchical dimension, Point points to the associated node in the next layer;
(3) The data contained in each leaf node can exist in the form of a binary [Key: Value], Key represents the value of the N-th dimension, and Value represents the value of TID-List corresponding to a complete path;
(4) Each node contains a keyword with a special value of All, represented by "*" or null, which represents the situation when the key of the node is an arbitrary value;
(5) All keys on the path from the root node to a leaf node constitute a record in the fact table, and the path length from the root node to all leaf nodes is equal.

Definition 2.2 Prefix Expansion. In the date cube, the elements of one aggregation unit may be partially the same as the elements of another aggregation unit. If a dimension value or a combination of multiple dimension values exist in the prefix of the aggregation unit, a common prefix is formed. Units with a common prefix are stored only once in the dwarf cube structure.

Definition 2.3 Suffix Coalescing. In the date cube, there are multiple different aggregation units, but the same aggregate values can be obtained, because these different aggregation units are aggregated from the same tuple in the data table, as shown in the dwarf cube structure, different paths from the root node can reach the leaf node with the same aggregation value. In this case, leaf nodes with the same aggregation value are merged, stored only once, and multiple paths share the same suffix.

The structure of the dwarf cube corresponding to the rainfall fact Table 1 is shown in Fig. 1.

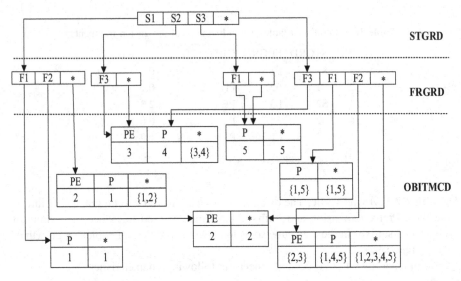

Fig. 1. Dwarf cube corresponding to the rainfall fact table

2.2 Improved Dwarf Cube

Through the process of prefix expansion and suffix merging, dwarf cube removes part of the redundancy in the cube and achieves the purpose of compressing the volume. However, the study found that the structure of dwarf cube is complex and there is still some redundancy. Therefore, an improved dwarf cube structure is proposed to compress the cube volume to a greater extent.

For sparse cubes with high dimensions and large amounts of data, there is some redundancy when generating aggregation units. In the Fig. 1, the node in the second layer on the path <S3, F1, P> contains only "F1" and "*", where the "*" item is redundant, removing the "*" item does not affect the query result. Similarly, for leaf nodes that contain only one measurement value and item "*", only the measurement value needs to be stored. The improved dwarf cube structure is shown in Fig. 2.

The improved dwarf cube gets the same aggregated result as the original dwarf cube when querying. Although some intermediate nodes and leaf nodes remove the "*" item, they can still respond to queries with "*". When a node contains only a dimension value and a "*" value, the node can match both its own dimension value and the "*" value. In sparse high-dimensional data set, there will be many nodes with only a dimension value and "*" value. The improved dwarf cube further compresses data storage space without affecting query efficiency by reducing redundant "*" items.

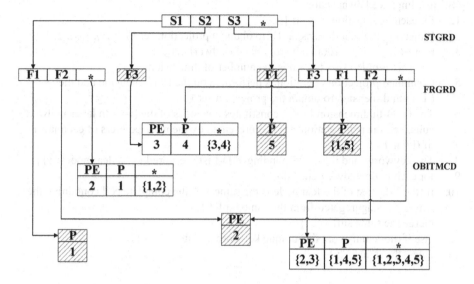

Fig. 2. Improved dwarf cube corresponding to the rainfall fact table

Algorithm Idea of Improved Dwarf Cube: Projection calculation is performed on the first dimension of the aggregation unit in the complete cube to obtain multiple dimensional values without repetition. The combination of these dimensional values constitutes the first layer of dwarf; Then the combination of the first and the second dimension is projected to obtain multiple combination values without repetition. The second layer of dwarf is formed by the projection of the second dimension. The first and the second layers are connected by related Pointers to represent the combination order. And so on, each layer is connected by Pointers to form the structure of directed acyclic graph. The leaf node is composed of the dimension value of the last dimension and the corresponding TID. For multiple nodes with the same TID, if these units are clustered from the same (or more) tuples, the leaf nodes are suffixed. Except the root node, for nodes with only a dimension value and "*" value in each layer, the item "*" can be removed and only the single dimension value can be retained due to redundancy caused by the same aggregation value. The improved dwarf cube generation algorithm is described as below.

Algorithm 1. Improved dwarf cube generation algorithm
Input: A fact table with n non-hierarchical dimensions $\{D_1, D_2, ..., D_n\}$.
Output: Improved dwarf cube.
1.   for each aggregation unit do {
2.   Projecting the first dimension D_1 to obtain a projection set $V_1 = \{v_1, v_2, ...\}$;
3.   root $= V_1$;   // root is the root node of dwarf cube
4.   for (i=1; i<n; i++) do {   // n is the number of dimensions
5.   Combined projection operation is performed on the i-th dimension and the
     (i+1)-th dimension to obtain the projection set V_{i+1}.
6.   The (i+1)-th dimension in V_{i+1} constitutes the nodes of the (i+1)-th layer in dwarf
     cube, and the path relationship is represented by the node pointers of each layer;
7.   if (i = n-1) {
8.   The keywords and the corresponding TID-List are stored in the leaf nodes; }}}
9.   for each node in dwarf cube do {
10. if (the TID-List of the leaf node is the same && the complete cell containing the
     leaf node is aggregated from the same tuple) {
11. merge the same suffixes; }
12. else if (node only contains unique keyword and item "*") {
13. delete the item "*"; }}

The cube with D non-hierarchical dimensional attributes is divided into dimensional segments of size F, with a total of D/F segments. Construct each dimension fragment according to the improved dwarf algorithm, and the complexity is: $O((D/F) * n^2) = O(n^2)$, as the number of non-hierarchical dimensions increases, the dwarf cube's high compression performance is highlighted.

2.3   SPD-Cube

Map/Reduce [16] is based on the Hadoop Distributed File System (HDFS) [17]. It is a distributed computing framework, model and platform for parallel processing of large-scale data sets.

For the non-hierarchical dwarf cube with higher dimensions, although the dwarf structure has reduced large redundancy and storage space compared with the traditional cube construction and storage method, it is still hard to avoid the sharp expansion of dwarf cube due to the large amount of data and high dimensions of cube, resulting in large size of dwarf cube and complex data cube structure. A feasible solution is to segment all non-hierarchical dimensions of high-dimensional according to the principle of disjoint, and generate K low-dimensional sub-cubes [18]. Then, combined with the improved dwarf idea and Map/Reduce architecture, each sub-cube is constructed, stored and queried in parallel, generating a number of sub-dwarfs, thus put forward the Segmented Parallel Dwarf Cube (SPD-Cube) [19].

The establishment of SPD-Cube can be divided into three stages: Map stage, Shuffle stage and Reduce stage. The schematic diagram of parallel construction of SPD-Cube is shown in Fig. 3, which is described as follows:

(1) In the Map stage, the DFS system reads the original data one by one and processes the original data into a series of <key-value> pairs through the function Map. The key is composed of the compound key <SID, tuple>, SID represents the fragment number, tuple represents the aggregation unit, and value is TID-List, which represents the list of index numbers. For example, in Fig. 3, for the original tuple (A1, B1, C1, TID), the Map stage divides the tuple into <key, value> pairs, "SID = 1" in the key represents the first group, tuple = (A1, B1, C1), value = TID list.

(2) In the Shuffle stage, group the <key, value> pairs generated in the Map stage, merge the pairs with the same fragment number into the same group, and then send data from the same group of to the same Reduce node for further processing. For example, in Fig. 3, tuples with the same SID are allocated to the same group and sent to the same Reduce node.

(3) In the Reduce stage, the data set of the same group obtained in the Shuffle phase is processed by the Reduce function, each Reduce node processes <key, value> pairs with the same SID. According to the improved dwarf generation algorithm, each group is built, each Reduce output result is a sub-dwarf, output form is <key = tuple, value = {TID1, TID2, ... TIDn}>, where the value is the union of TID of tuples.

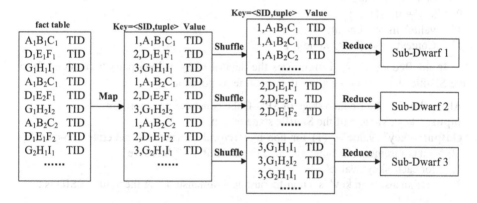

Fig. 3. SPD-Cube parallel construction diagram

3 Hydrological Data Cube Algorithm Based on SPD-Cube

The segmented parallel dwarf cube technology is proposed for non-hierarchical dimension. The idea of cube segmentation is adopted to divide the high-dimensional data into multiple low-dimensional segments. Then, by combining the idea of improved dwarf cube and Map/Reduce technology, dwarf cube was constructed for each low-dimensional segment in parallel, and several sub-dwarfs were generated to compress the cube volume and improve the query efficiency of MOLAP.

3.1 SPD-Cube Construction Algorithm

SPD-Cube parallel construction algorithm is divided into Map stage, Shuffle stage and Reduce stage. The Map stage inputs the original data set in the fact table in the form of

<key, value>, the key is composed of a compound key <SID, tuple>, SID is the fragment number, tuple is the aggregation unit, value is TID-List which represents the index number list.

Algorithm 2. The SPD-Cube construction algorithm
Input: <key, value>, key is tuple record, value is TID-List corresponding to tuple.
Output: <key', value'>:(1) for non-hierarchical dimensions: key'=<SID, tuple>, value'=TID-List; (2) for measure attribute: key'=<0, TID>, value'= measured value.
1.   The non-hierarchical dimensional cube $\{D_1, D_2, ..., D_n\}$ with n dimensions is divided into k low-dimensional segments according to the principle of non-intersection;
2.   for each <key, value> do {
3.   if (dimension in key is a non-hierarchical dimension) {
4.   for (i=0; i< number of tuples; i++) {
5.   key'=<SID, tuple>;   // group numbers are set for each dimension fragment
6.   value'=TID-List;
7.   Context. write(key', value'); }
8.   else if (dimension in key is a measurement dimension) then{   // generate table of TID_Measure
9.   key'=<0,TID>; }
10.   value'=measurement;
11.   Context. write(key', value'); }}

In the Reduce stage, it minimizes the data of <key, value> pairs transmitted from the Shuffle stage while keeping it as pristine as possible.

Algorithm 3. The Reduce stage of SPD-Cube construction algorithm
Input: < key', value'> of the Shuffle stage output.
Output: <key", value">. (1) for non-hierarchical dimensions: generate sub-dwarfs; (2) for measurement attribute: generate the table of TID_Measure.
1.   for each <key', value'> {
2.   if (dimension in key' is a non-hierarchical dimension && the value of SID is the same) {
3.   sub-dwarf is constructed for the tuples of the same group.   //according to the algorithm of constructing dwarf cube
4.   key"=tuple;
5.   value"=<TID_1, ..., TID_n>;
6.   Context. write (key", value");   }
7.   else if (SID=0) {
8.   key"=TID;
9.   value"=measurement;
10.   Context. write (key", value"); }}

3.2   SPD-Cube Query Algorithm

MOLAP query based on SPD-Cube can be divided into point query and range query. The point query is a simple traversal from the root node to the leaf node in depth-first mode, and matching along the chain of Pointers layer by layer according to the required query keywords until the leaf node. When the keyword is 'All', it represents matching item "*", the algorithm returns the TID-List of the leaf node. Finally, the corresponding measurement values of each TID index number are found in the TID_Measure table, and the final query result is obtained according to the given aggregation function.

The key to SPD-Cube point query is to decompose the dimensions involved in the query condition into corresponding non-hierarchical dimension segments to determine which dimensions are in the same sub-dwarf segment. Different query conditions correspond to different sub-dwarfs. Then look up the TID-List of the dimension attribute values from the sub-dwarf corresponding to each dimension segment, and perform an intersection operation on each TID-List to obtain the desired result.

In order to further improve the efficiency of SPD-Cube point query, the distributed parallel framework of Map/Reduce was used to conduct parallel point query on SPD-Cube. The parallel query diagram of SPD-Cube is shown in Fig. 4.

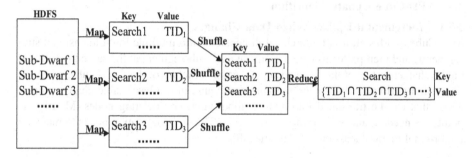

Fig. 4. SPD-Cube parallel query graph

Algorithm 4. The Map stage of SPD-Cube query algorithm
Input: (1) query conditions $Q = (v_1, v_2,...,v_n: M)$; (2) sub-dwarf collection{sub-dwarf$_1$,...sub-dwarf$_n$}; (3)the table of TID_Measure.
Output: <key', value'>: key 'is query condition Q, value' is TID-List.
1. divide the query conditions into sub-queries{Q_1, Q_2, ..., Q_n} according to the non-hierarchical dimension grouping principle.
2. for each sub-query {
3. find the sub-dwarf$_n$ corresponding to Q_n;
4. Using the point query method, the depth-first traversal is performed to match dimension value on   the sub-dwarf$_n$;
5. if (match successfully) {
6. Obtain the TID-List$_n$ corresponding to the dimension value;
7. else return (null); }
8. key' = Q_n;
9. value' = TID-List$_n$;
10. Context. write (key', value'); }

Algorithm 5. The Reduce stage of SPD-Cube query algorithm
Input: < key', value'> of the Shuffle stage output.
Output: <key", value">: key" is the query condition Q, value" is an aggregate value.
1.   for each <key', value'> {
2.   key"=Q;
3.   value"=TID-List1∩TID-List2∩TID-List3∩...;
4.   take the measurement value corresponding to each TID from the TID_Measure table, and get the result according to aggregate function.
5.   value"=result;
6.   Context. write (key", value"); }

3.3   SPD-Cube Update Algorithm

3.3.1   Incremental Update When Data Changes

SPD-Cube divides all non-hierarchical dimensions into multiple low-dimensional sub-segments, and then performs calculations on each sub-segment at the same time. Based on the characteristics of the directed acyclic graph structure of the dwarf cube, when the dimension attribute value or the measurement value in the basic table changes, operations that can be translated into adding, modifying, and deleting nodes. Modify the pointer as needed, and split or merge the nodes. The entire operation needs to maintain the original hierarchical semantic relationship.

(1) Modify

When the value of one (or more) dimension attribute(s) in the basic fact table changes, find the sub-fragment corresponding to the non-hierarchical dimension first, then the sub-dwarf of the sub-fragment update with node keys or split and merge nodes. Starting from the root node, look up the keywords on the corresponding layer from top to bottom along the chain of pointers. By cross-comparing the dimension value of the original node with the dimension value of the new tuple, the cells that do not need to be updated are scanned. If the new dimension does not exist in the original sub-dwarf, a new node needs to be created to store the new dimension value, and the pointer of the original node needs to be transferred. Cyclic updating other aggregation units that may be affected by the new tuple, including the keyword and TID-List, until all affected nodes have been updated.

When the value of one (or more) measure attribute in the basic fact table changes, the TID value corresponding to the original measure value is found according to the given condition, and then update the <TID, M> record in the TID_Measure table.

(2) Insert

When one (or more) new tuples are inserted into the basic fact table, for all non-hierarchical dimensions in the tuple, it is necessary to divide them into sub-fragments and then insert each sub-fragment into the corresponding sub-dwarf. Starting from the root node, loop over to determine whether the original sub-dwarf has the same dimension value as the new data unit, and performs pointer transfer and node merging or constructing new units as needed.

(3) Delete

When one (or more) tuple is deleted from the basic fact table, the relevant data for each sub-fragment should also be deleted in each sub-dwarf. First, find the path of the tuple fragment to be deleted in sub-dwarf. If the path is only a unique item, delete the node and the pointer of the unique item.

If there is more than one keyword in the passed node, and other tuples still need to use the keyword, only the relevant keyword in the leaf node will be deleted, that is, the last dimension value of the tuple fragment to be deleted, pointer transfer will be carried out as required, and then the TID-List of the relevant leaf node will be recalculated and updated. In addition, the <TID, M> corresponding to the tuple should be deleted from the table of TID_Measure.

3.3.2 Incremental Updates When a Dimension Changes

When non-hierarchical dimensions in the original fact table change, the corresponding sub-dwarf needs to be updated in a timely manner. In general, updates to non-hierarchical dimensions are classified as adding or removing one (or more) dimension to the fact table.

(1) When adding one (or more) non-hierarchical dimension in the basic fact table, if the number of newly added dimensions is greater than the maximum size of the current dimension fragment, the newly added dimensions shall be divided into segments, and a new sub-dwarf shall be generated from each segment separately, and it shall be dissected from other sub-dwarfs; If the number of newly added dimensions is less than the minimum size of the current dimension fragment, add the newly added dimensions to the currently smallest dimension fragment and recalculate the sub-dwarf.

(2) When one (or more) non-hierarchical dimension is deleted from the basic fact table, if the size of the deleted dimension fragment meets the size requirement of fragment division, the dimension attribute is deleted directly and the sub-dwarfs of the remaining dimension are recalculated. If the size of the deleted dimension fragment is less than the current minimum dimension fragment, the remaining dimensions in the dimension fragment are merged into other fragments, the sub-dwarf of the original fragment is deleted, and the sub-dwarf of the new fragment is recalculated.

4   Experiments and Performance Analysis

The experimental setup is a Hadoop cluster running on nine nodes, including one NameNode and eight DataNodes. In the cluster, the configuration of the NameNode is: CPU: Intel (R) Core (TM) i5-7200U CPU @2.50 GHz, RAM: 8G, HDD: 240G, software environment: Ubuntu Linux; development language: Java.

The experimental data were the basic data and daily rainfall data of some important hydrological stations in China, with a total data volume of $7.2 * 10^6$ records. A total of ten non-hierarchical dimensional attributes (grade of hydrometric station, grade of flood-reporting station, item of observation, year of station establishment, month of station establishment, year of withdrawal of station, month of withdrawal of station,

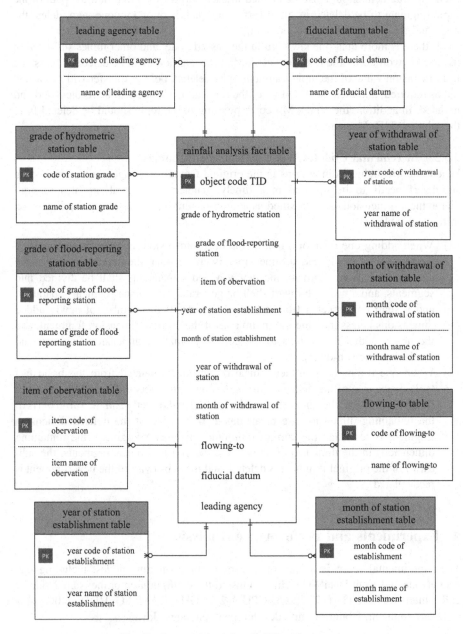

Fig. 5. Star model of the rainfall analysis topic

flowing-to, fiducial datum, leading agency) and one measurement attribute (precipitation) were selected. Before the experiment, the data was normalized, data fields were extracted from the original hydrological database, and the null value and inconsistent data were removed according to the data conversion and cleaning rules. Combining actual analysis requirements, the star model of the rainfall analysis topic is designed, as shown in Fig. 5.

In the case of different dimension numbers, the performance of SPD-Cube proposed in this paper is compared with the traditional dwarf method in terms of storage space and query efficiency to test whether the SPD-Cube method can reduce the storage space and query efficiency occupied by high-dimensional data cubes.

4.1 Comparison of SPD-Cube and Traditional Dwarf Cube

The performances of SPD-Cube and dwarf methods in storage space and time to respond to queries are compared on different non-hierarchical dimensions (six to ten). It can be seen from the experimental results in Fig. 6. As the number of non-hierarchical dimensions increases, the space occupied by the SPD-Cube method grows slowly, and the storage space required by the dwarf method increases exponentially. SPD-Cube reduces the space for storing redundant "*" items on basis of traditional dwarf. When the amount of data is larger, the more redundant items can be reduced, thus achieving the purpose of space compression. As can be seen from the experimental results in Fig. 7, with the increasing of number of dimensions, the query response time required by SPD-Cube method is much less than that by dwarf method, and the growth rate is slow. Therefore, SPD-Cube has some advantages in reducing storage space and improving query efficiency.

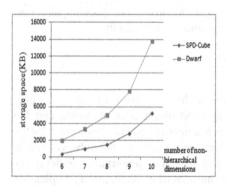

Fig. 6. Comparison of storage space for different number of dimensions

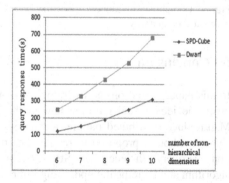

Fig. 7. Comparison of query response time for different number of dimensions

4.2   Comparison of SPD-Cube and Data Cube Based on Map/Reduce

The performances of SPD-Cube and original data cube based on Map/Reduce(MR-Cube) methods in storage space and time to respond to queries are compared on different amount of data ($3 * 10^6$ to $7 * 10^6$). It can be seen from the experimental results in Fig. 8 that the storage space of SPD-Cube is smaller than that of MR-Cube in the case of different data volumes, and with the increase of data volumes, the storage space difference between them is also larger. Because the traditional data cube uses a spatial cube structure to store data, every piece of data in the fact table is stored once. As can be seen from the experimental results in Fig. 9, the larger the query data volume, the gap of query response time between SPD-cube and MR-Cube is larger. MR-cube uses inefficient sequential search, and the directed acyclic-graph structure of SPD-Cube provides an efficient index mechanism. Therefore, SPD-Cube has some advantages in reducing storage space and improving query efficiency.

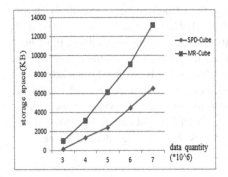

Fig. 8. Comparison of storage space for different data quantity

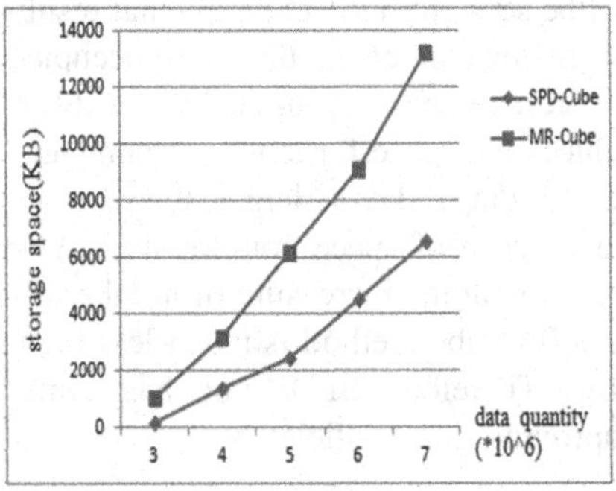

Fig. 9. Comparison of query response time for different data quantity

5   Conclusion

In this paper, based on the hydrological non-hierarchical dimension in the data cube, using the idea of cube fragmentation, the improved dwarf cube algorithm and the Map/Reduce distributed parallel technology, a segmented parallel dwarf cube (SPD-Cube) method is proposed to reduce data redundancy, compress the cube volume to improve query efficiency. The SPD-Cube parallel generation algorithm, parallel query algorithm and incremental update algorithm are introduced in detail, and a complete set of high-dimensional data cube calculation methods for non-hierarchical dimensions are formed. Experiments show that SPD-Cube method has certain advantages in reducing storage space and improving query response efficiency.

Of course, there are still many places that need to be researched and improved in this paper. For example, we can study how to use data cubes to efficiently improve query efficiency for the hierarchical dimension of hydrological data. In addition, we are also thinking about the application of SPD-Cube method in food industry, medical industry and other fields.

Acknowledgement. This work has been supported by the National Key Research and Development Program of China (No. 2018YFC1508106).

References

1. Xike, X., Xingjun, H., et al.: OLAP over probabilistic data cubes I: aggregating, materializing, and querying. In: 2016 IEEE 32nd International Conference on Data Engineering (ICDE), Helsinki, pp. 799–810 (2016)
2. Deqing, C., Wenyu, W., Haikun, Y.: Application research on data warehouse of hydrological data comprehensive analysis. In: 2010 3rd International Conference on Computer Science and Information Technology, Chengdu, pp. 140–143 (2010)
3. d'Orazio, L., Bimonte, S.: Multidimensional arrays for warehousing data on clouds. In: Hameurlain, A., Morvan, F., Tjoa, A.M. (eds.) Globe 2010. LNCS, vol. 6265, pp. 26–37. Springer, Heidelberg (2010). https://doi.org/10.1007/978-3-642-15108-8_3
4. Waiwen, J., Dongping, X., Xiaoxia, Z.: Research on MOLAP storage and query technology based on multidimensional database. Comput. Eng. Appl. **33**(24), 166–168 (2005)
5. Dong, X., Jiawei, H., Xiaolei, L., et al.: Computing iceberg cubes by top-down and bottom-up integration: the StarCubing approach. IEEE Trans. Knowl. Data Eng. **19**(1), 111–126 (2007)
6. Lingyun, W., Haining, L.: Implementation of dimensional aggregation with MOLAP layers. Comput. Eng. Des. **28**(19), 4595–4596+4715 (2007)
7. Lakshmanan, L.V.S, Russakovsky, A., Sashikanth, V.: What-if OLAP queries with changing dimensions. In: 2008 IEEE 24th International Conference on Data Engineering, Cancun, Mexico, pp. 1334–1336 (2008)
8. Daoguo, L., Duoqian, M.: Research and improvement of tree pruning algorithm. Comput. Eng. **31**(8), 19–21 (2005)
9. Xiuzhen, Z., Pauline, L.C., Guozhu, D.: Efficient computation of iceberg cubes by bounding aggregate functions. IEEE Trans. Knowl. Data Eng. **19**(7), 903–918 (2007)
10. Sheng'en, L., Shan, W.: Star Cube—an efficient data cube implementation method. J. Comput. Res. Dev. **9**(1), 587–593 (2004)
11. Zheng, S., Jiawei, H., Dong, X.: MM-Cubing: computing iceberg cubes by factorizing the lattice space. In: 16th International Conference on Scientific and Statistical Database Management, Santorini Island, Greece, pp. 213–222 (2004)
12. Zhiwei, N., Jinhua, M., Xuemin, M.: The calculation of a closed cube on a set. Comput. Eng. Appl. 36–38+58 (2011)
13. Weiji, X.: CC-Bitmaps: an effective index technology of the closed cube. South China University of Technology (2010)
14. Wei, W., Jianlin, F., Hongjun, L.: Condensed cube: an effective approach to reducing data cube size. In: Proceedings 18th International Conference on Data Engineering, San Jose, CA, USA, pp. 155–165 (2002)
15. Zhibing, S., Houkuan, H., Hongmin, L.: A compressed data cube structure that retains semantics. Comput. Eng. **34**(13), 37–39 (2008)
16. Shim, K.: MapReduce algorithms for big data analysis. In: Madaan, A., Kikuchi, S., Bhalla, S. (eds.) DNIS 2013. LNCS, vol. 7813, pp. 44–48. Springer, Heidelberg (2013). https://doi.org/10.1007/978-3-642-37134-9_3

17. Dean, J., Ghemawat, S.: MapReduce: simplified data processing on large clusters. Commun. ACM **51**(1), 107–113 (2008)
18. Leng, F., Bao, Y., Wang, D., Yu, G.: A clustered dwarf structure to speed up queries on data cubes. In: Song, I.Y., Eder, J., Nguyen, T.M. (eds.) DaWaK 2007. LNCS, vol. 4654, pp. 170–180. Springer, Heidelberg (2007). https://doi.org/10.1007/978-3-540-74553-2_16
19. Arnab, N., Cong, Y., Philip, B., Raghu, R.: Distributed cube materialization on holistic measures. In: Proceedings of the 27th International Conference on Data Engineering, Hannover, Germany, pp. 183–194 (2011)

A Machine Learning Approach to Prostate Cancer Risk Classification Through Use of RNA Sequencing Data

Matthew Casey[1]([⊠]) [iD], Baldwin Chen[1], Jonathan Zhou[2] [iD], and Nianjun Zhou[3]

[1] Ardsley High School, Ardsley, NY 10522, USA
mattcasey02@gmail.com, baldwinchen@gmail.com
[2] Horace Greeley High School, Chappaqua, NY 10514, USA
jozhou@students.ccsd.ws
[3] IBM, 1101 Kitchawan Road, Yorktown Heights, NY 10598, USA
jzhou@us.ibm.com

Abstract. Advancements in RNA sequencing technology have made genomic data acquired during sequencing more precise, making models fitted to sequencing data more practical. Previous studies conducted regarding prostate cancer diagnosis have been limited to microarray data, with limited successes. We utilized The Cancer Genome Atlas' (TCGA) prostate cancer sequencing data to test the viability of fitting machine learning models to RNA sequencing data. A major challenge associated with the sequencing data is its high dimensionality. In this research, we addressed two complementary tasks. The first was to identify genes most associated with potential cancer. We started by using the mutual information metric to identify the most significant genes. Furthermore, we applied the Recursive Feature Elimination (RFE) algorithm to reduce the number of genes needed to identify cancer. The second task was to create a classification model to separate potential high-risk patients from the healthy ones. For the second task, we combated the high dimensionality challenge with Principal Component Analysis (PCA). In addition to high dimensionality, another challenge is the imbalanced data set that has a 10:1 class imbalance of cancerous and healthy tissue respectively. To combat this problem, we used the Synthetic Minority Oversampling Technique (SMOTE) to create synthetic observations and equalize the class distribution. We trained and tested a logistic regression model using 5-fold cross-validation. The results were promising, significantly reducing the false negative rate as compared to current diagnostic techniques while still keeping the false positive rate low. The model showed great improvements over previous machine learning attempts to diagnose prostate cancer. Our model could be applied as part of the patient diagnosis pipeline, helping to improve accuracy.

Keywords: Machine learning · RNA sequencing · Prostate cancer diagnosis · Upsampling · Logistic regression · Multi-layer perceptron · Auto-encoding · Mutual information · Recursive feature elimination · The Cancer Genome Atlas

© Springer Nature Switzerland AG 2019
K. Chen et al. (Eds.): BigData 2019, LNCS 11514, pp. 65–79, 2019.
https://doi.org/10.1007/978-3-030-23551-2_5

1 Introduction

Prostate cancer is the second leading cause of cancer death among men in the United States, with an estimated 31,620 deaths predicted in 2019 [1]. The cancer is characterized by a malignant tumor found within the prostate and is mainly found in men 65 and older [1]. Currently, diagnosis begins with a preliminary blood test called a Prostate Specific Antigen (PSA) test. If the patient shows an abnormally high PSA level, a doctor may recommend that they undergo a second PSA test and a Digital Rectal Exam (DRE), in which a doctor palpates the prostate to check for abnormalities implying a tumor. A prostate biopsy may be recommended if a lump is detected or if PSA levels continue to rise[1]. After the biopsy, the cancer is assigned a Gleason score and stage, both of which indicate the severity of the cancer [2].

There is a great need for an accurate early detection method for prostate cancer. Various studies have raised concerns regarding PSA testing and the effects it has on patients. It is estimated that PSA tests have overdiagnosis rates of between 23% and 42% [3]. Often, this can lead to unnecessary anxiety and decreased general health for patients [4]. The results from the PSA tests often prompt patients to get a biopsy done; however, this can lead to more confusion due to the inaccuracy of biopsy-based diagnosis. A study done in 2013 found that the standard 12-core biopsy method of diagnosis is very ineffective [5]. According to the study, Gleason scores were underestimated in 47.8% of patients. In addition, they found that the false negative rate could be 30% or more, meaning that many patients with cancer were diagnosed as healthy. Lastly, they found the detection rate was even lower for patients with lower PSA values. These issues have prompted researchers to seek new methods of diagnosis, one of which is through the use of RNA sequencing data.

RNA sequencing is a gene sequencing technique which gives a more precise view of a cell's transcriptome than previous microarray or Sanger sequencing based methods [6]. DNA microarrays are used to measure gene expression levels; however, they are not as detailed as RNA sequencing is [6]. The data acquired from RNA sequencing is important for cancer classification because certain differences in the transcriptome can indicate the presence of prostate cancer. The sequencing could be performed after the biopsy in addition to slide pathology.

A recent study done by the National Institute of Health compiled RNA sequencing data for 33 different types of cancer. The database, called The Cancer Genome Atlas (TCGA) contains data for both healthy and cancerous samples for each cancer. We selected prostate cancer because of its prevalence in society and current problems with diagnosis. In addition, the prostate cancer dataset has a relatively high number of samples. All of the datasets have many more cancerous samples than healthy samples and have a sample-feature imbalance. This means that the number of features significantly outnumbers the samples, which is common in gene and healthcare research. Thus, picking one of the cancers with a relatively high amount of samples ensures that the results are more robust.

---

[1] https://www.cancer.gov/types/prostate/psa-fact-sheet#q1.

The purpose of this project was two-fold. The first was to identify genes related to cancer. The second was to produce effective classification models that can outperform standard biopsies and microarray-based models. The costs of genomic sequencing have decreased, making prostate cancer classification using RNA sequencing data a much more practical option to enhance diagnostic techniques [7]. Biopsies do not always result in accurate results, and our model can enhance the results of biopsies, leading to greater predictive accuracy.

The paper is organized as follows. The second section will discuss related works. The third section will discuss problem formulation and data acquisition. The fourth section will address how to identify key gene sequences related to Prostate Cancer using mutual information and recursive feature elimination. In the fifth section, we will develop a binary classification model, utilizing logistic regression to distinguish cancerous and non-cancerous individuals. In the sixth section, we will discuss additional efforts we explored or are currently in progress. In the final section, we will summarize what we have accomplished and future works.

2   Related Work

Various attempts have been made to classify cancer based on tissue samples. These various studies have used microarray, clinical, imaging, and RNA sequencing data. Many recent works have utilized microarray datasets for cancer classification. The most recent study developed a new approach which aimed to improve accuracy when using microarray data for classification [8]. Early studies conducted that attempted to diagnose prostate cancer with machine learning utilized microarray datasets. Various studies were conducted using different methods and were tested on five different microarray datasets [9–14]. These studies aimed to predict whether a cancer would metastasize or not. Although the results of all the microarray based studies are significant, the increased information gain about the transcriptome from RNA sequencing should give way to improved classification accuracy.

A recent study done attempted to diagnose prostate cancer with machine learning through use of clinical data [15]. They trained an Artificial Neural Network (ANN) with data consisting of 22 clinical features. They found that although their model performed well, it needed improvements before being suitable for clinical applications.

The TCGA database has already been used for cancer classification. The data contained in the database goes beyond RNA sequencing data. A recent study used slide images for classification of lung cell cancer through use of a Convolutional Neural Network (CNN) [16]. Their results improved upon those achieved by doctors in manual diagnosis.

Based on our research, no published studies have yet attempted using prostate cancer RNA sequencing data from the TCGA database for cancer classification. However, studies have been published using the breast cancer dataset for classification [17, 18]. The two main challenges associated with all of the TCGA datasets for advanced analytical study are high dimensionality and class imbalance. The high dimensionality poses two problems. Firstly, there are so many features that the models will not be able to accurately separate the data into healthy and sick. Thus, the results of any model trained on the data may be poor. Secondly, the number of features is many times greater than the number of samples. This is known as a feature-sample imbalance, and it causes models trained on the data to be unstable and leads to overfitting. This makes the results of any model trained on the data unreliable. The studies conducted by Danaee et al. [17] and Golcuk et al. [18] using TCGA breast cancer tested many methods to combat these problems.

Danaee et al. tried various methods of dimensionality reduction such as a Stacked Denoising Autoencoder (SDAE), differentially expressed genes, PCA, and KPCA. They tried each of these methods with three different models, an Artificial Neural Network (ANN), a Support-Vector Machine (SVM) with a linear kernel, and an SVM with a radial basis function kernel (SVM-RBF). They calculated five metrics for each model: accuracy, sensitivity, specificity, precision, and f-measure. They found that the highest accuracy was attained using the SDAE for dimensionality reduction followed by the SVM-RBF model. This method also had the highest F-measure. The highest sensitivity was achieved with the SDAE as well, but with the ANN model. The KPCA with SVM-RBF model attained the highest specificity and precision.

Golcuk et al. conducted a study which aimed to improve upon the results achieved by Danaee et al. As a baseline they tried three dimensionality reduction algorithms (PCA, KPCA, and NMF), followed by a SVM. They also tried utilizing a ladder network, which does not require a reduction in dimensionality. They found that the ladder network slightly outperformed both the SDAE and SVM models from the previous study in almost all metrics. The only metric in which it performed worse was specificity, showing a slight decrease as compared with the KPCA and SVM-RBF model.

Only one of these studies however dealt with the class imbalance problem. The first study utilized the same SMOTE technique that we will use in this study to increase the number of samples in the dataset. The second study, however, failed to address the class imbalance. Both their test and validation sets have only 20% of the data, and with such a low number of healthy samples in the dataset already, each of these datasets had very few healthy samples. As a result, the results of the model could have been inflated and makes their results less reliable than those of the first study.

Both of these studies used only the gene expression data and neglected to use the other three datasets obtained from RNA sequencing. The other datasets present a significant source of information that could help to improve the performance of models.

3   Data Acquisition and Preprocessing

In this section, we will discuss how we acquire the genomic datasets from the TCGA database and the preprocessing method to combine the datasets based on shared IDs. Furthermore, we will boost our dataset using the Synthetic Minority Oversampling Technique (SMOTE). The purpose of doing the preprocessing is to prepare the data for use in the gene selection and classification models, and to combat model instability issues.

3.1   Data Acquisition

In order to download the RNA sequencing data from the TCGA database, we used an open-source tool called TCGA-Assembler 2 [19]. All four main RNA sequencing datasets were downloaded as well as the clinical data. Various types of genomic data were included: Exon expression, Exon Junctions, Isoform expression, and Gene expression data. Exons, or sections of genes that provide the code needed to create proteins, as opposed to introns, which are designed to not code for anything. Exon data in the TCGA data set displays the positions of the exons on the individual chromosomes based off of distance from the ends of the chromosomes. It also shows the expression levels of each of the exons. Exon Junctions, or the positions where two exons meet, show mutations occurring when the individual exons are combined to form a single pre-mRNA chain and allow for scientists to observe similarities in specific mutations in a specific area that are common to all cancerous patients [8]. Isoforms, genes that serve almost identical purposes, but are composed of different exons in different orders or completely different bases, show the genes that are similar to each other. Lastly, gene expression data shows the exact levels at which a gene is expressed, allowing researchers to identify genes that are common to cancerous patients and those that are common to healthy individuals.

After downloading the data, the built-in processing functions were used to clean up the raw data. These functions extract the most useful parts of the data for analysis. For instance, for gene expression data, normalized count values are extracted, and for exon expression data, RPKM (Reads Per Kilobase of transcript, per Million mapped reads) values are extracted. These values are selected because they are comparable from sample to sample, unlike the raw data, making them far more useful for analysis. Each dataset was outputted in a tab-delimited text file and was used later in our own preprocessing. At this point, the high dimensionality of the data becomes very clear, with nearly 600,000 total features across the four datasets (Table 1).

Table 1. Number of features for each dataset

	Gene	Exon Junction	Isoform	Exon
Feature count	20531	249566	73598	239321

3.2   Preprocessing of RNA Sequencing Data

The four different data types arising from the RNA sequencing were preprocessed separately due to slight differences in structure. For each data type, the data was first formatted so that the index was the sample ID and the columns were the features. Then, the samples were categorized as either 0 for cancerous or 1 for cancer free, by extracting the 13th and 14th digits of the sample ID. A few samples were originally categorized as 6, or metastatic, and those were reclassified to be 0.

For the gene quantification and isoform quantification datasets, each gene/isoform had two columns, one corresponding to the raw count per transcript and the other corresponding to a scaled value which was independent of transcript length. In order to make more accurate comparisons of gene and isoform expression between samples, the raw count column was dropped for each gene/isoform. In addition, the scaled values were multiplied by one million to convert them into a Transcripts Per Million (TPM) value. Following completion of individual preprocessing, the data were merged into one larger dataset. Since all of the data came from the same sequencing process, each sample had data in each dataset, and the merge was done using the sample ID as the reference.

Table 2. Sample gene data

Patient ID	UNK-100130426	UNK-100133144	UNK-100134869	UNK-10357	UNK-10431
TCGA-2A-A8VL-01	0.0	1.294659	0.788167	11.358769	85.259114
TCGA-2A-A8VO-01	0.0	1.121938	0.593362	8.164714	52.753502

Due to space restrictions, we display only a snapshot of one of the four datasets (Table 2). This dataset contains 20531 genes and 546 patients, however we only show five genes and two patients. The values in the table represent the expression of the gene.

3.3 Upsampling with Synthetic Minority Oversampling Technique

Before any testing could be done using machine learning models, we first upsampled the data in order to deal with model instability resulting from the 10:1 class imbalance between cancerous and normal (minority) cells. The Synthetic Minority Oversampling Technique (SMOTE) is a method in which synthetic data of the minority class is created which closely resembles the original data [20]. In order to do this, it creates a new observation randomly on the imaginary line connecting an existing data point with the data point closest to it. As a result, the data stays in the same general cluster, but the amount of samples is increased. We increased the number of samples from 496 cancerous and 50 healthy to 3000 cancerous and 1500 healthy. By doing this, we increased the robustness of the results of the classifiers since more healthy samples are contained in each test part of the test/train split. Without this method, there was a very low amount of healthy samples in the test set, which may have artificially inflated the accuracy.

4 Key Gene Sequence Identification

In this section, we will first identify a set of genes related to Prostate Cancer using the techniques of mutual information and Recursive Feature Elimination. The mutual information is used to give a preliminary measure of each gene's importance. The RFE is then used to minimize the gene sequences needed to perform diagnostics and predictions.

4.1   Gene Sequence Selection Using Mutual Information

The first step in the feature selection process was the identification of genes that are significant in determining whether a cell is cancerous. We used the mutual information metric to determine the most significant genes. Mutual information is a metric which quantifies the amount of information gained about one variable by observing the other. High mutual information between a gene and the target would mean that knowing the expression of the gene would give the model a good indication of whether the sample is cancerous. Equation 1 defines the mutual information between two discrete variables X and Y and was used to calculate the mutual information between each gene and the target (cancerous or healthy).

$$I(X;Y) = \sum_{y \in \mathcal{Y}} \sum_{x \in \mathcal{X}} p(x,y) \log\left(\frac{p(x,y)}{p(x)p(y)}\right) \tag{1}$$

The results of the mutual information are summarized in Table 3.

Table 3. Mutual information rankings

Gene ID	Mutual information
Gene_POLR2H-5437	0.167398
Gene_GSTM4-2948	0.163962
Gene_APOBEC3C-27350	0.161822
Gene_HPN-3249	0.159367
Gene_ETNK2-55224	0.154814
Gene_ANGPT1-284	0.154056
Gene_GSTP1-2950	0.153113
Gene_LURAP1-541468	0.151971
Gene_TMLHE-55217	0.151118
Gene_MCF2-4168	0.149987
Gene_SLC19A1-6573	0.149467
Gene_NKX2-3-159296	0.148963
Gene_PYCR1-5831	0.148119
Gene_PLP2-5355	0.146773
Gene_HOXC6-3223	0.145437
Gene_EFNB1-1947	0.144863
Gene_NKAPL-222698	0.144755
Gene_MARCKSL1-65108	0.14449
Gene_ASPA-443	0.144207
Gene_NECAB1-64168	0.143234

Table 3 contains the top twenty genes with the highest mutual information values. The meaning of these genes can be found on the National Center for Biotechnology Information (NCBI) database. We examined several genes from Table 3 using the NCBI database and noticed some of those genes are theorized to be related with cancers.

4.2 Feature Selection with Recursive Feature Elimination

In order to improve the performance of a model trained on the selected features, feature selection through Recursive Feature Elimination (RFE) was employed on the gene quantification data. The advantage of this method over purely using the genes with the highest mutual information is that it will pick genes that are not related to each other. Two genes may have very high mutual information values but if they are highly correlated to each other, dependent on the same hidden variable, or are otherwise related, one of them is redundant when doing classification. With RFE, the chosen genes are not related, and therefore the variance of the data is better represented. Using RFE, the highest ranked genes selected will give the highest accuracy when classifying a sample. A further advantage of this method is that because only a few genes need to be used when doing classification, the large sample/feature imbalance no longer exists, and any models created will be more robust.

We trained the model on the genes which had a mutual information value of above 0.05. We selected this threshold to ensure we maximize the number of significant genes used in the selection model. With this threshold, we kept only 2565 genes for use from the original 20532 genes (Fig. 1).

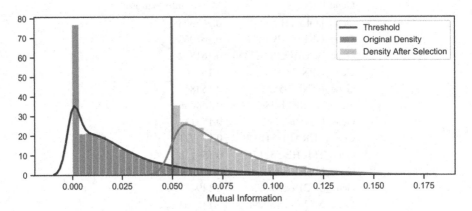

Fig. 1. Density plot of mutual information w/threshold of MI = 0.05

Without such initial filtering, the RFE is infeasible from a computational perspective. During each iteration of training, a logistic regression classifier was trained to classify samples as healthy or cancerous. After training, the model ranked the features in order of importance, and the least important feature was removed. This process was repeated until there was only one feature left, and a ranked list of all of the genes was obtained. The results of the RFE are summarized in Table 4.

Table 4. Recursive feature elimination rankings

Gene ID	Mutual information	RFE rank (Logistic)
Gene_HPN-3249	0.159367224	1
Gene_GSTM1-2944	0.134699717	2
Gene_APOE-348	0.059651147	3
Gene_MRPL41-64975	0.091810873	4
Gene_TRIB1-10221	0.060442517	5
Gene_RPL18A-6142	0.051059319	6
Gene_ISG15-9636	0.052118627	7
Gene_RHOB-388	0.056353286	8
Gene_AMACR-23600	0.107861574	9
Gene_MYLK-4638	0.134553865	10
Gene_FLNA-2316	0.058148017	11
Gene_EEF1G-1937	0.059776923	12
Gene_RPL37-6167	0.074615666	13
Gene_WFDC2-10406	0.099545078	14
Gene_APOC1-341	0.078565194	15
Gene_RPLP0-6175	0.054809254	16
Gene_PCP4-5121	0.082975965	17
Gene_GDF15-9518	0.058465024	18
Gene_RPL28-6158	0.103847129	19
Gene_ATP5MF-9551	0.055716636	20

Many of the genes found in Table 4 are not found in Table 3. This indicates that many of the genes with high mutual information were related to each other and as a result were not found to be as significant by the RFE.

Compared to the model that will be generated in Sect. 5 using PCA, any models using this set of features will have higher interpretability. These features are genes that can be easily related to pathology. In addition, it is economically advantageous to sequence only a few genes as compared with sequencing the whole genome [21]. Besides the topics discussed in Sect. 6, in the future we would like to apply RFE to the combined data, and use the resulting features for classification.

5   Predictive Modeling

In this section, we developed a predictive model using the combined data. The features were condensed using PCA (Principal Component Analysis), allowing us to develop a classification model. In our current research, a logistic regression model was developed to fit the preprocessed datasets from Sect. 3. In Sect. 6, we will discuss how to extend this model to include the prediction of cancer stages using more advanced modeling approaches.

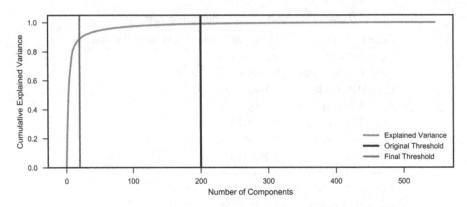

Fig. 2. PCA cumulative explained variance

5.1   Dimensionality Reduction with Principal Component Analysis

High dimensionality prevents models from producing effective results because it is difficult for the model to extract target information from the data. Dimensionality must be reduced in order for models to function. Besides RFE, the other dimensionality reduction technique we used was Principal Component Analysis (PCA). This technique reduces a set of possibly correlated variables into a set of principal components, which are uncorrelated variables. These principal components represent the variance in the data, with each successive principal component representing less variance.

In order to pick how many components to reduce the data to, we found the cumulative explained variance for each number of components (Fig. 2). We originally selected 200 components for the combined data and 50 for each individual dataset. We found however that the logistic regression models trained on this data were heavily overfitting, achieving nearly 100% accuracies. A full summary of the performance of the combined data 200 principal component model can be found in Table 5. We hypothesized that the overfitting was caused by the feature sample imbalance, and that there were still too many features being used. To combat this problem, we decided to reduce the number of principal components for each data type to 5 and the number on the combined data to 20. We chose this number of components because they explained nearly 90% of the variance, a heuristic often utilized to determine how many principal components to use. Each of these five resulting datasets was later used for classification.

5.2   Fitting of the Models

We created six different logistic regression models, one for each of the post-PCA datasets, and one for the top five features from the recursive feature elimination. We chose logistic regression due to the linear nature of the data and the robustness of the logistic regression algorithm. 5-fold cross-validation was performed and repeated twenty times in order to validate the robustness of the classifier. The upsampled data was used, allowing us to do the cross-validation, which would otherwise not have been possible.

Various metrics were calculated for each model trained during the cross-validation, and the final metric represents the average of all of these results. The metrics collected were accuracy, precision, recall, f1 score, ROC AUC, PR AUC, false-negative rate, and false-positive rate. Precision quantifies how many instances that were predicted positive are actually positive. Recall quantifies how many instances that are actually positive were predicted as such. F1 score is the harmonic average of precision and recall and it quantifies the overall performance of the model. ROC (Receiver Operating Characteristics) AUC is the area under the curve with false positive rate on the x-axis and true positive rate on the y-axis. PR (Precision-Recall) AUC is the area under the curve with recall on the x-axis and precision on the y-axis. Accuracy was chosen just to see a benchmark, but due to the imbalanced nature of the dataset, the precision, recall, and f1 score provide a much more meaningful quantification of the model's performance. ROC AUC was calculated since it is a standard metric. However, PR AUC gives a more meaningful representation of the model performance. Figure 3 shows the PR and ROC curves for the combined PCA model.

Table 5. Model metrics

	Dataset	Accuracy	Precision	Recall	F1 score	ROC AUC	PR AUC	FN	FP
PCA	Gene	0.850	0.783	0.736	0.759	0.913	0.841	0.084	0.065
	Exon	0.896	0.859	0.806	0.832	0.952	0.902	0.062	0.042
	Exon Junction	0.919	0.885	0.859	0.872	0.970	0.946	0.045	0.036
	Isoform	0.879	0.832	0.780	0.805	0.933	0.887	0.070	0.050
	Combined	0.972	0.969	0.942	0.955	0.995	0.987	0.019	0.010
	Combined200	0.999	0.995	1.000	0.998	0.999	0.999	0.000	0.001
RFE	Gene	0.961	0.926	0.941	0.912	0.993	0.985	0.014	0.024

Table 5 gives a summary of each model's performance. We created seven models and report eight metrics on each of them. For PCA models, we applied the PCA algorithm to five different prostate cancer sequencing datasets (Gene, Exon, Exon Junction, Isoform, and all the former combined) and trained a logistic regression model on the resulting principal components. These models are referred to as PCA gene, PCA exon, PCA exon junction, PCA isoform, and PCA combined respectively. The PCA Combined200 model uses the top 200 principal components instead of the top 20, as discussed in Sect. 5.1. The RFE gene model uses features selected by the recursive feature elimination algorithm applied to the Gene dataset.

Out of the four individual PCA models, the gene model performed the worst. However, the RFE gene model outperformed all the individual PCA models, implying that RFE is a better way to do feature reduction, as there may be less information loss. This gives another reason for RFE to be applied to the combined data in future work.

The combined PCA model performed the best on all metrics except false negative rate, with a ROC AUC score of 0.99, PR AUC of 0.98, and an F1 score of 0.96. The RFE gene model achieved a false negative rate lower than the combined PCA model. However, it performed more poorly with regards to PR AUC and F1 score. Thus, overall the RFE gene model is a less effective classifier than the combined PCA Model.

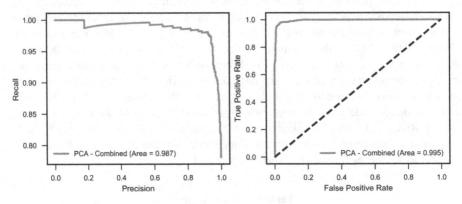

Fig. 3. PR (left) and ROC (right) curves for combined PCA model

6 Discussion

Our current approach, logistic regression, is limited due to its inability to determine the stage of cancer because it is a binary model. In order to predict the patient's cancer stage, we attempted using a multi-classification approach. We utilized an Artificial Neural Network (ANN), a deep learning approach, with a Rectified Linear Unit (ReLU) as our activation function in the middle layers and SoftMax in the final layer. We used the post-PCA combined dataset to train the model.

To analyze the performance of our ANN, we compared the performance (cross-entropy) of the model to that of a zero model. The zero model is a model that represents randomly guessing which stage of cancer (if any) the patient had. We found that our model made no significant improvement in performance as compared with the zero model. The zero model had an accuracy of 0.2501 and our model had an accuracy of 0.2839. We hypothesize that this may be because the stage and size of the cancer are not reflected by the divergence in RNA sequences from healthy cells. Our next step is to incorporate the clinical data to create a more effective classifier.

During our analysis we found that any model generated on our dataset was highly dependent on input data and thus unstable in terms of its performance with regards to the test-train split made. This is because any stable model must have a stable set of statistical properties in the training set and a reliable statistical relationship between the features and target. However, for such a small sample size, as is the case here, any split would significantly disturb the distribution of feature values of the population in training and testing data sets, especially for testing set. This is because slight changes in the number of healthy tissues in the testing set would drastically change the generated model performance. Thus, our challenge becomes whether we can build a stable model with this data set.

A potential area for improvement is in the area of feature reduction. Our current approach with PCA may be limited by its linearity, and this may be improved upon using an autoencoder technique to create and highlight new features of importance. One of our on-going efforts is to use multiple layers of auto-encoding neural networks to do the compression. However, we must be cautious of problems associated with model complexity in this case.

A limitation which was brought up concerned the performance of mutual information with RFE. It was suggested that using a random forest in place of RFE may be more optimal as it already has a built-in information theoretic feature selection criterion and may improve overall performance. The reasoning is that since no features are completely correlated, there is still loss of information when the RFE is used. In other words, if one feature is removed, there may be an inadvertent and undesirable loss of information. After creating a random forest of 300 trees with depths of 2 we found that the most important selected features were comparable to those as selected by mutual information. For example, we found that PLOR2H-5437, which ranked first in mutual information, was ranked seventh in the random forest.

7 Conclusion and Future Work

Our model significantly improves upon previous prostate cancer research done using microarray data, and research done using RNA sequencing breast cancer data. In addition, it improves upon the accuracy of the standard biopsy procedure and has the potential to standardize the process. The current method of identifying cancer through the use of a biopsy is inaccurate and based on the skill of the technician, which varies from location to location [22]. Our model provides a way to improve upon this, because it can match or exceed the accuracy of diagnosis of a highly experienced technician, and removes the disparity that exists between highly experienced and less experienced technicians. In addition, there is an estimated 30% false negative rate associated with the current method of diagnosis [5]. Our method greatly reduces this number to 1.3%, meaning that many fewer cases of prostate cancer are missed, especially the less severe cases. Although our proposed system still requires the use of a biopsy, it has the potential to be a very useful tool which can help doctors accurately diagnose patients and minimize false positives and false negatives.

In this paper, we have presented an approach which has the potential to improve upon the microarray method of prostate cancer identification greatly. Our accuracy also significantly improved upon previous attempts to classify prostate cancer using machine learning. Our method, however, only demonstrates the initial advantages of using machine learning in cancer prediction. This work is limited by the dataset used, which forces us to create false observations to fix the class imbalance. Future work needs to address this class imbalance and attempt other ways of dealing with it which may have less of a potential to artificially inflate the results. Currently, our model does not utilize clinical data when making a classification. In future work, we will also fit the clinical data to see what effect on model effectiveness that data may have. We are testing deep learning approaches to compare this approach to traditional machine learning methods. We would also like to apply the RFE algorithm to the combined data to see how it performs. We hypothesize that it would outperform the results achieved with the PCA.

References

1. American Cancer Society Cancer Facts & Figures 2019. https://www.cancer.org/content/dam/cancer-org/research/cancer-facts-and-statistics/annual-cancer-facts-and-figures/2019/cancer-facts-and-figures-2019.pdf. Accessed 27 Jan 2019
2. American Joint Committee on Cancer: AJCC Cancer Staging Manual, 8th edn. Springer, New York (2017). https://doi.org/10.1007/978-1-4757-3656-4
3. Draisma, G., et al.: Lead time and overdiagnosis in prostate-specific antigen screening: importance of methods and context. J. Natl. Cancer Inst. **101**(6), 374–383 (2009)
4. Albertsen, P.C.: The unintended burden of increased prostate cancer detection associated with prostate cancer screening and diagnosis. Urology **75**(2), 399–405 (2010)
5. Serefoglu, E.C., Altinova, S., Ugras, N.S., Akıncıoğlu, E., Asil, E., Balbay, M.D.: How reliable is 12-core prostate biopsy procedure in the detection of prostate cancer? Can. Urol. Assoc. J. **7**(5–6), E293–E298 (2013)
6. Kukurba, K.R., Montgomery, S.B.: RNA sequencing and analysis. Cold Spring Harb. Protoc. **2015**(11), 951–969 (2015)
7. Deep Learning for genomic data analysis. https://repositorio-aberto.up.pt/bitstream/10216/106492/2/205645.pdf. Accessed 27 Jan 2019
8. Mitra, S., Saha, S., Acharya, S.: Fusion of stability and multi-objective optimization for solving cancer tissue classification problem. Expert Syst. Appl. **113**, 377–396 (2018)
9. Penney, K.L., et al.: mRNA expression signature of Gleason grade predicts lethal prostate cancer. J. Clin. Oncol. **29**(17), 2391–2396 (2011)
10. Cuzick, J., et al.: Prognostic value of an RNA expression signature derived from cell cycle proliferation genes in patients with prostate cancer: a retrospective study. Lancet Oncol. **12**(3), 245–255 (2011)
11. Erho, N., et al.: Discovery and validation of a prostate cancer genomic classifier that predicts early metastasis following radical prostatectomy. PLoS ONE **8**(6), e66855 (2013)
12. Mo, F., et al.: Stromal gene expression is predictive for metastatic primary prostate cancer. Eur. Urol. **73**(4), 524–532 (2018)
13. Tyekucheva, S., et al.: Stromal and epithelial transcriptional map of initiation progression and metastatic potential of human prostate cancer. Nat. Commun. **8**(1), 420 (2017)
14. Sharifi-Noghabi, H., et al.: Deep Genomic Signature for early metastasis prediction in prostate cancer. bioRxiv, 276055 (2018)
15. Takeuchi, T., Hattori-Kato, M., Okuno, Y., Iwai, S., Mikami, K.: Prediction of prostate cancer by deep learning with multilayer artificial neural network. bioRxiv, 291609 (2018)
16. Coudray, N., et al.: Classification and mutation prediction from non-small cell lung cancer histopathology images using deep learning. Nat. Med. **24**(10), 1559 (2018)
17. Danaee, P., Ghaeini, R., Hendrix, D.A.: A deep learning approach for cancer detection and relevant gene identification. In: Pacific Symposium on Biocomputing, vol. 22, pp. 219–229 (2016)
18. Golcuk, G., Tuncel, M.A., Canakoglu, A.: Exploiting ladder networks for gene expression classification. In: Rojas, I., Ortuño, F. (eds.) IWBBIO 2018. LNCS, vol. 10813, pp. 270–278. Springer, Cham (2018). https://doi.org/10.1007/978-3-319-78723-7_23

19. Wei, L., Jin, Z., Yang, S., Xu, Y., Zhu, Y., Ji, Y.: TCGA-Assembler 2: software pipeline for retrieval and processing of TCGA/CPTAC data. Bioinformatics **34**(9), 1615–1617 (2017)
20. Chawla, N.V., Bowyer, K.W., Hall, L.O., Kegelmeyer, W.P.: SMOTE: synthetic minority over-sampling technique. J. Artif. Intell. Res. **16**, 321–357 (2002)
21. Advances in the molecular diagnosis of cancer. https://repositorio.unican.es/xmlui/bitstream/handle/10902/14278/Cadrecha Sanchez Natalia.pdf?sequence=1&isAllowed=y. Accessed 27 Jan 2019
22. Cancer Classification using Gene Expression Data with Deep Learning. https://www.politesi.polimi.it/bitstream/10589/138427/7/thesis.pdf. Accessed 27 Jan 2019

User Level Multi-feed Weighted Topic Embeddings for Studying Network Interaction in Twitter

Pujan Paudel, Amartya Hatua$^{(\boxtimes)}$, Trung T. Nguyen$^{(\boxtimes)}$,
and Andrew H. Sung$^{(\boxtimes)}$

School of Computing Sciences and Computer Engineering,
The University of Southern Mississippi, Hattiesburg, MS 39406, USA
{pujan.paudel,amartya.hatua,trung.nguyen,andrew.sung}@usm.edu

Abstract. Over half a billion tweets on a wide range of topics are posted daily by hundreds of millions of Twitter users. Insights of user behavior and network interactions can be applied to practical applications like targeted advertising, viral marketing, political campaigns, etc. In this paper, we propose a Multi-Feed Weighted Topic Embeddings (MFWTE) model to study user network interaction and topic diffusion patterns on Twitter. Our method extracts topic embeddings from multiple views of a Twitter user feed and weights them according to their content authoring roles, where the authored tweets, replied tweets, retweeted tweets, and favorited tweets are the views we separate for constructing the embeddings. We test the proposed method using two different topic modeling algorithms: (i) Latent Dirichlet Allocation (ii) Twitter-Latent Dirichlet Allocation. The users in our study are divided into multiple hierarchies based on their activity composition regarding individual topics, and the effectiveness of MFWTE is evaluated in the multi-hierarchical setting. The performance of our method on friendship recommendation and retweet behavior prediction task is evaluated using various ranked retrieval measures. The results indicate that our MFWTE method for topic modeling of Twitter users improves over various previous baselines. We conclude our work by applying the proposed model, MFWTE to discover various information diffusion patterns on Twitter.

1 Introduction

Microblogging platforms, such as Twitter and Reddit, have emerged as the primary platform on the internet for users all around the world to engage in discussions over a large variety of topics. Users can use Twitter to associate any tweet with any number of hashtags which allows the platform to aggregate large volumes of related tweets in real time. Hashtags in Twitter serve as an important explicit clustering mechanism that helps to recommend users and topics of similar interest in the platform. However, the use of hashtags as a primary mechanism for topic detection or any other interaction study would be inconsistent and highly effective across Twitter users: In a previous study, Boyd et al. [9]

© Springer Nature Switzerland AG 2019
K. Chen et al. (Eds.): BigData 2019, LNCS 11514, pp. 80–94, 2019.
https://doi.org/10.1007/978-3-030-23551-2_6

have shown that only 5% of tweets contain a hashtag. Also, a group of users might interact about a topic using a set of different hashtags. Studying link formation and information diffusion pattern of users and their tweets within the wider Twitter network using topic models open a wide area of applications for targeting advertisements, user recommendations, and network analysis.

The most popular topic model is Latent Dirichlet Allocation (LDA) [6] which has grown quite popular for modeling large document collections with large text lengths. Traditional LDA models fail when applied over short and imbalanced texts with relatively shorter document lengths and skewed topic-word distribution. To limit the discrepancy of LDA models over short texts, various solutions have been proposed. Weng et al. [1] and Hong et al. [2] tried aggregating all the tweets of a user as a single document to account for short document length. Zuo et al. [3] performed modeling of distributions over topics using word co-occurrence matrix to alleviate the problem for short texts. Zhao et al. [4] proposed a model, Twitter-LDA by assuming the association of a single tweet with a single topic.

Embedding vectors have been used successfully in Natural Language Processing tasks such as sentiment detection in Tang et al. [7], and Text Classification and Neural Translation in Zou et al. [8]. Compared to generating embedding vector for texts, generating embeddings for multi-view data is non-trivial as it requires dealing with different modalities and distributional properties of the views. Benton et al. [5] proposed a weighted variant of Generalized Canonical Correlation Analysis to learn multiview embeddings of Twitter users. We were motivated by the effectiveness of their method in learning embeddings for the tasks of user engagement prediction, friendship prediction, and demographic attribute prediction.

User recommendation problem in Twitter space has been investigated before using various approaches. Graph-based approaches such as in Armentano et al. [11] consider topological position of users in network graph (followers as well as friends) to recommend potential followers. In the same work, they also propose content-based recommendation by comparing tweet content of the users social graph. Garcia et al. [12] propose weighted content-based recommendation method by identifying popularity, activity, location, mutual friends and tweet contents as features. Golder et al. [13] investigates structural approaches for user recommendation by using features like reciprocity, shared interests, shared audience and filtered people. Twittomender, a recommender system proposed by Hannon et al. [14] used an ensemble of profiling strategies, both content-based, and collaborative-filtering based, to recommend user profiles to follow. Experiments have shown that a combination of collaborative methods is more precise than individual content-based methods [15]. Similarly, sentiment-based [17] approach has also been applied for Twitter user recommendation.

Pennacchiotti and Gurumurthy [18] investigated topic models for social media user recommendations using an adapted user level LDA model, replacing documents with users' Twitter stream. Their LDA system significantly outperformed the TF-IDF representations of users' tweets, demonstrating the applicability of topic models for capturing user-level behavior. Similarly, Ramage et al., in [19] used Labeled-LDA [20] to characterize Twitter users using topic-models. They demonstrated the weighted combination of TF-IDF model of user tweets and Labeled-LDA together performing well in the task of user-recommendation.

Most of the work done on retweet link prediction use similarity matching between the source profile and target profile of a user, or matching score between re-tweetable tweets under study and users topic of Interest. Multiple approaches have been adopted to create the user profiles of a user for similarity matching. Xu and Yang [22] proposed TF-IDF based bag of word profiles for each user on their similarity based model for retweet prediction. The work by [23] studied the information sharing strategies of users in online social networks under the strategical features of Interest Matching, Linguistics, Information Trustability and Information Freshness. The evaluation of strength of features in the same study reported the TF-IDF weighted Bag Of Words (BOW) similarity of reposted tweets performing as the most powerful signal on predicting user retweet behavior. They also compared the optimal information strategy model with LDA topic models and reported LDA models outperforming TF-IDF models using retweet strategy as features.

Most of the work related to topic modeling in the Twitter environment has focused on inferring topic distributions of individual isolated tweets, deviating from user topic profiling view-point. While [18] and [19] are the closest to our work, the scale of their evaluation sizes is relatively small, as they use sample size 8 and 10 respectively for positive/negative test users for evaluating experimental results. The evaluation methods used by our study use Information Retrieval metrics like Precision @K, Recall @K, Mean Reciprocal Rank (MRR) in comparison to traditional machine learning metrics, like accuracy and ROC curve used in previous works. The behavior of topic models and their effectiveness on network analysis on a much larger evaluation size needs to be explored more for its possible application in Big Data systems. The objectives of this work are as follows: (1) Comparing the performance of different topic modeling algorithms on user-level distributions for the task of network analysis, (2) Studying the effects of dividing Twitter feed into multiple views, based upon their content authoring source, (3) Evaluating the advantages of using weighted embeddings over non-weighted topic embeddings, (4) Dividing the study users into multiple hierarchies based on their topical activity and studying the effect of weighted topic embeddings on them, (5) Discovering information Diffusion patterns in user networks through the Multi-Feed Weighted topic models.

The paper is organized as follows: Sect. 2 discusses the data collected and methods applied in our work; Sect. 3 covers the experimental setup for our research; Sect. 4 highlights our results on the weighted multi-feed topic embeddings for friendship recommendation and retweet link prediction; Sect. 5 investigates the findings and discusses contribution of the research for practical applications; Sect. 6 concludes our work.

Table 1. Topics and hashtags used

Topic	Hashtags
Animal rights	#animalrights, #animalabuse, #huntingkills, #saveanimals
Domestic violence	#metoo, #domesticviolence, #sexualviolence, #violenceagainstwomen
Book lovers	#bookworm, #books, #booklover, #bibilophile, #amreading
Net neutrality	#netneutrality, #savetheinternet
Mental health	#anxiety, #mentalIllness, #depression, #mentalhealth, #suicideawareness

2 Methodology

This section describes our methodology. We begin with the description of the dataset that we collected, the pre-processing pipeline, followed by the different views and embeddings of Twitter feed that we formulated and the topic modeling strategies that were applied.

2.1 Description of Dataset

Twitter lacks an explicit concept of topic space on their system. We compiled an individual topic as a collection of related hashtags. For our study, we selected five random topics and extracted five most common hashtags related to these topics, by observing the tweets associated with the users who tweeted multiple-tweets around those topics. The topics and their related hashtags used in our study are presented in Table 1.

We downloaded 2000 users each who tweeted at least more than 10 tweets under any single topic described in Table 1. While downloading the users, users with verified accounts and users with non-English profiles were removed. For each user, we collected 400 of their most recent tweets. We applied various pre-processing steps on these tweets to improve the quality of topics learnt by the models. We removed emoticons and other special characters which are a common source of noise on Twitter data. We removed all tweets not authored in English language, low-frequency words, stop words, HTML tags, and URLs from the tweets. We converted our entire vocabulary to lower cases, lemmatized them and expanded common English contraction words. Most importantly, we removed all the hashtags from the tweet texts, to ensure that the hashtags wont influence the learned topic distributions.

For network analysis of users friendship and retweet graph using the learned topic embeddings, we expanded social links of users in our study by downloading user information of 1500 of their followers and 1500 of their friends. Again, we collected 400 of their most recent tweets and passed them through the same pre-processing pipeline as described above.

2.2   Topic Modeling

We applied two different topic modeling algorithms on the tweets of users passed through pre-processing pipeline: traditional LDA model and Twitter-LDA model. From here on, we refer to traditional LDA model as Canonical LDA. We used the implementation of [6] for Canonical LDA while [4] for Twitter-LDA. We subjected both of the topic models under identical parameters of Dirichlet prior for Document-Topic distribution (α), and Dirichlet prior for Topic-Word Distribution (β), of 0.5 and 0.01 respectively. The number of topics was set to 6, one more than the number of actual topical classes, to account for the background class inferred by Twitter-LDA. For both of the cases, Gibbs sampling was applied for model parameter estimation. We planned on using WNTM [3], but due to extreme memory consumed by the word occurrence matrix of WNTM, we were unable to apply this model to our study.

2.3   Multi-feed Topic Modeling

To capture powerful representations of Topic Embeddings, we broke down Twitter feeds of individual users into multiple views based on their content authored sources: (A) **Authored View:** View composed with the tweets primarily authored by the users. (B) **Replied View:** View composed with tweets sent as a reply to other tweets. (C) **Retweeted View:** View composed with tweets forwarded by the user. (D) **Favs View:** Views composed with tweets favorited (liked) by the user. We are aware of community detection techniques using such multi-view approaches of data for community identification before. Greene and Cunningham [21] construct a heterogeneous collection of content-based views, incorporating views like tweet content, list text, mentions, retweets to produce unified graph representations for the task of community detection. Kwak et al. [10] compare trend analysis of users on their large-scale work of Twitter by observing the behavior of trending topics in 'Singleton', 'Reply', 'Mention' and 'Retweet' views. A similar methodology of differentiation between content source of tweets was done in the work of [16] where the comparison model selected profiling strategy of a user as either an 'author' or a 'retweeter' using topic similarity scores based on past tweets and retweets. Our work differs from the work of [16] in that the authors separated tweet source as a noise removal strategy for behavior prediction, while we are incorporating multiple views to capture dynamic content creation and sharing mechanisms of a Twitter user, extending beyond hard profile limitations of an 'author' or a 'retweeter' and allowing multiple content diffusion roles to be studied. We ran the same set of topic detection algorithms with identical parameters as discussed in Sect. 2.2 on the Multi-Feed topic models. For rest of the work, we refer the topic models of unseparated, traditional twitter feed as **single-feed** embeddings while the separated feed as discussed before as **multi-feed** embeddings.

2.4   Multi-feed Weighted Topic Modeling

We felt the need to weigh the multi-feed topic embeddings of our user feeds discussed in Sect. 2.3 with different weightage in order to produce efficient retrival results. We used the implementation by [5] for generating weighted embeddings of our multi-feed topic models, to propose our final model, MFWTE. We explored multiple view lengths {15, 20, 40, 100}, multiple view weightages {2, 5, 10, 20, 40, 80} and performed grid search over the model to investigate the effect of multi-view weighted topic embeddings.

2.5   Hierarchical Study of Twitter Users

We divided the users in our study into three hierarchies based on a user's Twitter activity towards topical distributions. We used results of the same topic modeling algorithm used in Sect. 2.2 to extract the hierarchies of users. The hierarchies of users were defined as follows: (A) **Tier 1 Users:** are the primary content creator for a topic, who tweet 85–95% about an individual topic. (B) **Tier 2 Users:** are secondary content creators for a topic, who compose relatively lesser tweets about a topic while their topic composition being 70–85% on an individual topic. (C) **Tier 3 Users:** very rarely author tweets by themselves related to a topic. Their Twitter activities amount to 50–60% about an individual topic. The users who fall under this tier generally have a multitude of interests and tweet almost equally about multiple topics. A similar distribution of study users into multiple buckets can be observed in [18], where they divided their experimental setup into head, torso and tail for investigating topic models for social media recommendation.

Table 2. Distribution of user tiers and evaluation set size per user for friendship recommendation

Evaluation Tier	Friend sample	Non-friend sample	Dev-set
1st Tier	530	530	1534
2nd Tier	344	344	1120
3rd Tier	419	419	1463

Table 3. Distribution of user tiers and evaluation set size per user for retweet behavior

Evaluation Tier	Retweet sample	Non-retweet sample	Dev-set
1st Tier	123	250	880
2nd Tier	106	220	1090
3rd Tier	150	289	1200

The formulation of our tiers differs from their division of buckets as their method divided users according to the number of followers, while our division is based on the topical activity of the users. The motive behind dividing the user base into multiple tiers was to study the effect of topic modeling on network prediction tasks separately on these multiple levels, as we observed the behavior of the topic models, and multiple views of it, vary amongst users who have different content authoring behavior over the platform.

3 Experimental Setup

This section explains the framework of analysis that we used to evaluate our method. We selected the task of friendship recommendation and retweet link prediction to examine the performance of topic models on users network interaction. Following relationship and retweeting relationship have been proven to be closely correlated with users interest as reported by Weng et al. [1].

For evaluating our models on the task of friendship recommendation, we made our evaluation sets equally balanced on all our tests by including an equal count of friend (positive) samples and non-friend (negative) samples. Every user has a set of friend and non-friend users. Non-friend users are defined as the users who are followed by ten of another user's friends, but who are not followed by the user. Compared to distinguishing between friend/non-friends of a user, identifying non-retweet links between Twitter users is non trivial. If user A is following user B, and user A has retweeted more than 10 tweets from B, then we sampled a directed link from A to B under positive class. If user A is friends with user B, but has less than 10 retweet links coming from user B and more than 10 friends who have 10 or more retweet links from user B, user B is sampled under the negative set. For retweet behavior prediction tasks, we followed prior adopted approaches by previous works, [23] to sample nearly double non-retweet (negative) samples compared to retweet (positive) samples. For both of the link prediction tasks, we isolated the Dev-set used to learn the multi-view embeddings from those that were used in the evaluation sets, to remove the possibilities of biases on the learnt embeddings For every tier of users, we held out random 100 target users; for each target user, we selected a positive set of users and a negative set of users. The distribution of evaluation set size for individual user, tier-wise, is displayed in Tables 2 and 3 for the two link prediction tasks, respectively. As seen in the tables, our evaluation size is much larger compared to those of previous studies. Scoring was done by ranking the compared cosine distance between the topic vectors of target users and users of the evaluation set. Evaluation was done using ranked retrieval metrics of Precision @K, Recall @K, as well as Mean Reciprocal Rank (MRR). A friendship connection between Twitter users is scored as a positive "hit" for the task of friendship recommendation while a directed retweet link between two users is scored as a positive "hit" for the task of retweet link prediction. The size of our evaluation sets for different studies as well as for different tiers of users was different, so we modified the metrics of Precision @K and Recall @K slightly to account for the percentage of positive class size present in the evaluation set. For example, Recall @0.5 refers to Recall @50% of the count of positive samples present in the evaluation set.

4   Results

This section discusses the results of our experimental study. We begin with the comparison of single-feed and multi-feed topic embeddings followed by the performance of different topic detection algorithms. We report the performance of weighted multi-feed topic embeddings. This section is concluded by the results on the application of weighted multi-feed embeddings to discover various topic diffusion patterns.

4.1   Comparison of Multi-feed with Single Feed Twitter Topic Embeddings

Before comparing the performance of multi-feed and single-feed topic model using ranked retrieval methods, we did an initial inspection of the quality of topic vectors using machine learning classifiers and machine learning evaluation metrics. For our proposed testing framework of friendship recommendation as discussed in Sect. 3, we trained a Support Vector Machine (SVM) using linear kernels for multi-feed and single-feed topic embeddings. We used the results from the topic modeling algorithms discussed in Sect. 3 as the input feature vectors. Evaluation was done using 10-fold Leave One Out cross-validation. For both of the link prediction tasks of friendship recommendation and retweet prediction, the multi-feed topic embeddings reports higher values of F1 score and accuracy over the baseline single-feed embeddings in all of the three hierarchies of users. The comparison tables of F1 score and accuracy is not displayed for the sake of brevity.

Next, we evaluated our Multi-Feed Embeddings using Ranked Information Retrieval measures. For the experimental framework of friendship recommendation, we compared Multi-Feed Topic Embeddings with the baseline of Single-Feed Topic Embeddings and TF-IDF embeddings of user tweets. Figure 1 shows the recall @K curve as a function of number of recommendations. Multi-Feed approach outperforms the Single-Feed approach as well as the TF-IDF Embeddings while evaluating under Precision @K, Recall @K and Mean Reciprocal Rank (MRR). This holds true for all the tiers of users we had divided, so we learned that dividing the tweets in Multi-Feed views, based upon their activity source, helps in improving the topic models for user behavior. We repeated identical Ranked Retrieval evaluation for retweet link prediction using topic models by comparing Multi-Feed Topic Embedding against multiple baselines of Single-Feed Topic Embedding, Topic Embedding of retweets collection of a user, and Bag Of Words (BOW) Weighted TF-IDF embeddings of a users tweet. The topic embeddings performed better than TF-IDF embeddings as expected. But, as seen in Fig. 1, neither the Multi-Feed Embeddings or Retweets topic Embeddings had a clear advantage in performance over the other as the value of Recall @K fluctuated with varying value of K.

We investigated two different types of topic models: Canonical LDA and Twitter LDA. Our analysis of the two modeling algorithms on the task of friendship recommendation shows that Twitter LDA outperforms the Canonical LDA

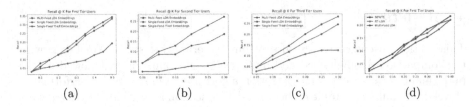

Fig. 1. Recall @K for friendship recommendation in (a) First Tier user, (b) Second Tier user, (c) Third Tier user, (d) Recall @K for retweet prediction

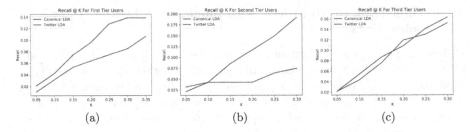

Fig. 2. Recall @K for canonical LDA and Twitter LDA (a) First Tier user (b) Second Tier user (c) Third Tier user

model under Recall @K metric for all the tiers of our user division. Figure 2 shows the Recall @K curve for the two topic models as a function of number of recommendations for all three tiers of users. There were identical results reported in the task of Retweet behavior prediction. The figures are left out for sake of brevity. In [18], the experimental results concluded that the user-level topic models are effective over tweet-level topic models. Though an indirect comparison under different dataset and different user-level tweet model (Twitter-LDA compared to their Labeled-LDA), our results disagree with their findings.

4.2   Comparison of Multi-feed Weighted Topic Embeddings with Non-weighted Embeddings

After verifying our initial proposal for the topic embeddings extracted from multiple content authoring sources in Sect. 4.1, our next question was if subjecting those learnt embeddings to weighted model would provide even more performance boost. Weighing models are important because of the different content authoring activity exhibited on different tiers of users and across different views of user feeds we had formulated. We did grid search over multiple weights and their combinations over multiple views to find the optimal weights for different tiers of users. Another example where Multi-View Embeddings have been used before was by Benton et al. [5] where they used multiple views (Ego Tweet, Friend tweets, Followers tweet) for the task of friendship prediction. For our evaluation, we used the best performing Multi-Feed topic embeddings from Sect. 4.1 and TF-IDF weighted Bag Of Words (BOW) representation of the multiple-content

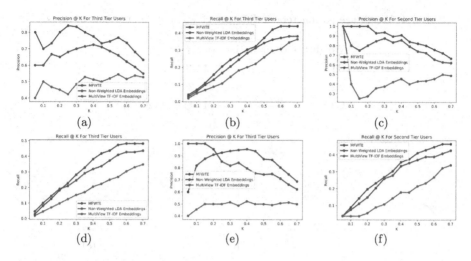

Fig. 3. Friendship recommendation comparison of MFWTE with baselines a & b (First Tier), c & d (Second Tier), e & f (Third Tier)

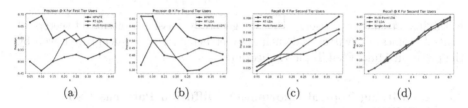

Fig. 4. Retweet prediction comparison of MFWTE with baselines a & b (First Tier), c & d (Second Tier)

sources tweets (Authored tweets, Replied tweets, Retweeted tweets and Favorited tweets) as discussed above as baselines. We performed identical grid searches of weights for the TD-IDF embedding baseline used for comparison. Comparison of Recall @K and Precision @K performance as function of number of recommendations is given in Fig. 3. We observed that the Multi-Feed Weighted Topic Embeddings (MFWTE) outperforms the best performing model from Sect. 4.1 and Multi-Feed TF-IDF Embeddings under Recall @K Metric. This verifies our proposed idea of Multi-Feed Weighted Topic Embeddings having an advantage over non-weighted embeddings.

We evaluated MFWTE for the task of retweet link prediction using identical experimental settings and similar baselines of best Multi-Feed Topic Embeddings from Sect. 4.1. For this study, we added Topic Embeddings of Retweeted Tweets and Multi-Feed TF-IDF embeddings as our baselines. It can be observed from the comparative analysis of the Precision @K and Recall @K curves in Fig. 4 that the MFWTE outperforms all other baselines for retweet link Prediction in the first and second tier of users. For the third tier of users, the Retweets Topic Embeddings perform the best. This can be explained by the topic composition

Table 4. Weightage parameters of different views on different Tiers of users

Weightage combination	α	β	γ	θ
C1	1	40	1	1
C2	20	20	1	1
C3	1	1	20	20

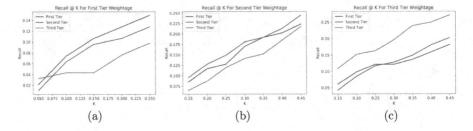

    (a)   (b)   (c)

Fig. 5. Recall values of MFWTE under (a) C1 (b) C2 (c) C3

instability of third tier users, who are generally passive content retweeters of multiple topics in Twitter and their retweet behavior goes uncaptured even by the powerful Multi-Feed Topic Embeddings.

4.3 Identifying Topical Information Diffusion Patterns Using MFWTE

In Sect. 4.2, we showed that weighing the embeddings improved the efficiency of our learnt topic models. We were motivated to find topical information diffusion patterns in the different hierarchy of users using the weighted embeddings. We applied MFWTE in investigating parameter combination space of topic embeddings at different content authoring behaviors throughout the different tiers of users. With α being the weight for **Authored View**, β being the weight for **Replied View**, γ being the weight for **Retweeted View**, and θ being the weight for **Favs View**, the weight combinations we used for the study of three different topical diffusion patterns is depicted in Table 4.

For Tier 1 of users, we proposed the primary content creators would engage in replying to topical tweets at a much higher rate than the other two tiers of users. Weightage combination C1 highlights our diffusion pattern for this case study. We subjected the test set of the other two tiers under same weight combination, which showed that C1 and the performance on number of positive friendship recommendation drops with the decrease in tier level. This validates our formulation of weight in Replied Views more for the first tiers of users. This is explained by the observation that non-topical users do not engage much in replying tweets related to a topic. We repeated similar experiment on Tier 2 of users with weight combination C2, proposing they engage in authoring the Tweet and Replying Tweets almost equally, while still being active in terms of authoring contents related to a topic compared to forwarding them. Weightage combination C2 highlights our diffusion pattern for this case study. We evaluated the performance of other tiers on the similar weight C2 and noted that the highest recall for positive friendship recommendation is observed for second tier users the most, followed by the first-tier users, which confirmed our initial assumption of weight attribution towards the second-tier users. Similarly, Tier 3 were content distributors; users who created less content for the topics themselves, but retweeted and favorited topical tweets. Weightage combination C3 highlights our diffusion pattern for this case study. When we subjected the three tiers of users under the weight combination of C3, it was observed that positive friendship recommendation performance in the third tier of users perform exceptionally well while the performance drops for the other two tiers with increase in tier level. The Recall @K curve of different information diffusion patterns and the effectiveness of MFWTE in capturing them is shown by Fig. 5. Thus, with the application of MFWTE using case specific weighted embeddings, we were able to demonstrate different content topic diffusion patterns for the different tiers.

5 Discussion

Users in the Twitter platform demonstrate highly complex activity of interaction, thus their authoring, replying and forwarding behavior in tweets are highly significant to determine their topics of interests and possible friendship connections. We divided a single stream feed of users into multi-view feeds, based upon their content authoring sources and were able to build better topic models for predicting friendship links and modeling retweet behavior in Twitter. The idea of segregating tweets with this configuration allows us to capture information diffusion in a highly dynamic environment like Twitter and create efficient collaborative filtering methods for user recommendations and information propagation. The four views which we have formulated can be extended to any number of activity sources, also extending to multiple social identities of an individual user (like Facebook) to build quality topic models. We applied two different variants of topic modeling algorithm on our datasets and discovered that Twitter-LDA has higher performance in the task of friendship recommendation than the LDA model designed for traditional documents. Thus, Twitter-LDA can be pursued as more efficient topic-modeling algorithm in other Big-Data analysis tasks for Twitter.

All of our Multi-Feed Topic Embeddings (both weighted as well as unweighted) perform better than their TF-IDF baselines. This result agrees with the findings by [18] where their adapted LDA system outperformed the TF-IDF baseline with statistical significance. Our results reinforce the claim that topic models are indeed good representations of user-level interests by demonstrating their efficiency in two link prediction tasks. The efficient performance of our final proposed model under Big-Data scale Information-Retrieval evaluation metrics (Ranked Retrieval when compared to ROC Curve used in [18]) confirms the application of Multi-Feed Weighted topic models as good representation of user level interests.

Improving over the Multi-Feed topic embeddings, our final model, MFWTE demonstrates even better results in all tiers of the user base we had formulated, opening up a wide and interesting area of application in high-impact marketing campaigns. MFWTE allowed learning highly dynamic topic embeddings based upon the tiers of user we are interested in targeting, as well as to capture different activity variance of Twitter users over multiple modes of interaction. Learning dynamic weights will help for improving targeted information outreach among different types of user bases. The topical information diffusion patterns we studied using MFWTE can be extended to analyze the spreading behavior of viral topics over different forms of interaction in a platform, as well as across different tiers of users in the platform. One such possible use case of it is the weighted embeddings of "Favs" view that we have learnt from our models. They can help identify the topic interest of users who may not be an enthusiast on a topic in terms of authoring them, or forwarding them but who are latent observers of the activities related to that topic. This type of users, who are quite common in the platform, might be reached for marketing and information campaigns, which otherwise might have remained unnoticed.

6   Conclusions

The main contribution of our paper is a weighted, multi-feed topic embedding which better captures topical interests of a users tweets and demonstrates better performance in discovering friendship connection and modeling retweet behavior on a large scale Twitter user network than previous models.

Our proposed methodology of segregating the user feed into multiple views based upon their content authoring sources helped us to capture the dynamics of the complex Twitter system and build better topic embeddings than traditional Twitter topic models. Further validation of the effectiveness of our model was done by evaluating them using different topic modeling algorithms and under different tiers of users. Being motivated by the effectiveness of multi-feed embeddings, we learned weighting parameters of the embeddings by grid-search over the parameter combination space improved over the non-weighted topic models. Our proposed final model Multi-Feed Weighted Topic Embeddings performs the best in Ranked Retrieval experiments, taking advantage of multi-view feeds as well as weighted embeddings learnt from WGCCA at the same time. Finally, by applying the MFWTE model on different tiers of our user-sets, we discovered multiple topic-level content authoring patterns of the users.

For future work, the multi-feed weighted models can be tested on other variants of topic modeling algorithms, like [3] and [19] requiring an extensive amount of working memory. Examining the effectiveness of these learnt topic models to community detection and other network analysis problems is another vital direction for future work.

References

1. Weng, J., Lim, E.P., Jiang, J., He, Q.: TwitterRank: finding topic-sensitive influential twitterers. In: Proceedings of the third ACM International Conference on Web Search and Data Mining, pp. 261–270. ACM, February 2010
2. Hong, L., Davison, B.D.: Empirical study of topic modeling in Twitter. In: Proceedings of the First Workshop on Social Media Analytics, pp. 80–88. ACM, July 2010
3. Zuo, Y., Zhao, J., Xu, K.: Word network topic model: a simple but general solution for short and imbalanced texts. Knowl. Inf. Syst. 48(2), 379–398 (2016)
4. Zhao, W.X., et al.: Comparing Twitter and traditional media using topic models. In: Clough, P., et al. (eds.) ECIR 2011. LNCS, vol. 6611, pp. 338–349. Springer, Heidelberg (2011). https://doi.org/10.1007/978-3-642-20161-5_34
5. Benton, A., Arora, R., Dredze, M.: Learning multiview embeddings of Twitter users. In: Proceedings of the 54th Annual Meeting of the Association for Computational Linguistics (Volume 2: Short Papers), vol. 2, pp. 14–19 (2016)
6. Blei, D.M., Ng, A.Y., Jordan, M.I.: Latent Dirichlet allocation. J. Mach. Learn. Res. 3(Jan), 993–1022 (2003)
7. Tang, D., Wei, F., Yang, N., Zhou, M., Liu, T., Qin, B.: Learning sentiment-specific word embedding for Twitter sentiment classification. In: Proceedings of the 52nd Annual Meeting of the Association for Computational Linguistics (Volume 1: Long Papers), vol. 1, pp. 1555–1565 (2014)
8. Zou, W.Y., Socher, R., Cer, D., Manning, C.D.: Bilingual word embeddings for phrase-based machine translation. In: Proceedings of the 2013 Conference on Empirical Methods in Natural Language Processing, pp. 1393–1398 (2013)
9. Boyd, D., Golder, S., Lotan, G.: Tweet, tweet, retweet: conversational aspects of retweeting on Twitter. In: 2010 43rd Hawaii International Conference on System Sciences (HICSS), pp. 1–10. IEEE, January 2010
10. Kwak, H., Lee, C., Park, H., Moon, S.: What is Twitter, a social network or a news media? In: Proceedings of the 19th International Conference on World Wide Web, pp. 591–600. ACM, April 2010
11. Armentano, M.G., Godoy, D., Amandi, A.A.: Recommending information sources to information seekers in Twitter. In: International Workshop on Social Web Mining, July 2011
12. Garcia-Gavilanes, R.O.G.G., Amatriain, X.: Weighted content based methods for recommending connections in online social networks (2010)
13. Golder, S.A., Yardi, S., Marwick, A., Boyd, D.: A structural approach to contact recommendations in online social networks. In: Workshop on Search in Social Media, SSM, July 2009
14. Hannon, J., McCarthy, K., Smyth, B.: Finding useful users on Twitter: twittomender the followee recommender. In: Clough, P., et al. (eds.) ECIR 2011. LNCS, vol. 6611, pp. 784–787. Springer, Heidelberg (2011). https://doi.org/10.1007/978-3-642-20161-5_94
15. Kywe, S.M., Lim, E.-P., Zhu, F.: A survey of recommender systems in Twitter. In: Aberer, K., Flache, A., Jager, W., Liu, L., Tang, J., Guéret, C. (eds.) SocInfo 2012. LNCS, vol. 7710, pp. 420–433. Springer, Heidelberg (2012). https://doi.org/10.1007/978-3-642-35386-4_31
16. Firdaus, S.N., Ding, C., Sadeghian, A.: Retweet prediction considering user's difference as an author and retweeter. In Proceedings of the 2016 IEEE/ACM International Conference on Advances in Social Networks Analysis and Mining (ASONAM 2016), pp. 852–859. IEEE Press, Piscataway (2016)

17. Gurini, D.F., Gasparetti, F., Micarelli, A., Sansonetti, G.: A sentiment-based approach to Twitter user recommendation. RSWeb@ RecSys, 1066 (2013)

18. Pennacchiotti, M., Gurumurthy, S.: Investigating topic models for social media user recommendation. In: Proceedings of the 20th International Conference Companion on World Wide Web, pp. 101–102. ACM, March 2011

19. Ramage, D., Dumais, S.T., Liebling, D.J.: Characterizing microblogs with topic models. ICWSM **10**(1), 16 (2010)

20. Ramage, D., Hall, D., Nallapati, R., Manning, C.D.: Labeled LDA: a supervised topic model for credit attribution in multi-labeled corpora. In: Proceedings of the 2009 Conference on Empirical Methods in Natural Language Processing: Volume 1-Volume 1, pp. 248–256. Association for Computational Linguistics, August 2009

21. Greene, D., Cunningham, P.: Producing a unified graph representation from multiple social network views. In: Proceedings of the 5th Annual ACM Web Science Conference, pp. 118–121. ACM, May 2013

22. Xu, Z., Yang, Q.: Analyzing user retweet behavior on Twitter. In: 2012 IEEE/ACM International Conference on Advances in Social Networks Analysis and Mining (ASONAM), pp. 46-50. IEEE, August 2012

23. Nguyen, D.A., Tan, S., Ramanathan, R., Yan, X.: Analyzing information sharing strategies of users in online social networks. In: Proceedings of the 2016 IEEE/ACM International Conference on Advances in Social Networks Analysis and Mining, pp. 247–254. IEEE Press, August 2016

Big Data Analytics for Nabbing Fraudulent Transactions in Taxation System

Priya Mehta[1], Jithin Mathews[1], Sandeep Kumar[3], K. Suryamukhi[1], Ch. Sobhan Babu[1(✉)], and S. V. Kasi Visweswara Rao[2]

[1] Indian Institute of Technology Hyderabad, Telangana, India
{cs15resch11007,cs15resch11004,cs17mtech01002,sobhan}@iith.ac.in
[2] Department of Commercial Taxes, Government of Telangana, Telangana, India
svkasivrao@gmail.com
[3] Plianto Technologies, Telangana, India
cs15mtech11017@iith.ac.in

Abstract. This paper explains an application of big data analytics to detect illegitimate transactions performed by fraudulent communities of people who are engaged in a notorious tax evasion practice called *circular trading*. We designed and implemented this technique for the commercial taxes department, government of Telangana, India. This problem is solved in two steps. In step one, the problem is formulated as detecting fraudulent communities in a social network, where the vertices correspond to dealers and edges correspond to sales transactions. In step two, specific type of cycles are removed from each fraudulent community, which were identified in step one, to detect the illegitimate transactions. We used *RHadoop* framework for implementing this technique.

Keywords: Data mining · Social network analysis · Big data · Goods and Services Tax · Fraud detection · Circular trading · Fraudulent transactions · Community detection

1 Introduction

Taxes are divided into two types namely, direct taxes and indirect taxes. The major difference between these two is the way in which they are collected. Direct taxes are collected from individuals and corporations. Income tax and gift tax are examples of direct taxes. Indirect taxes are imposed on the goods and services consumed. In this work, we work towards detecting evasion prevailing in the indirect taxation system. Value-added Tax (VAT) [26], and Goods and Services Tax (GST) [5] are indirect taxes. They are collected by a third party (*eg.*, shop keeper) from the consumer who purchases the goods. Finally, it is the consumer who would have to bear the burden of the tax payment.

Recent tax reforms in developing countries opted indirect taxation method to expand their tax base. Determining the "point of levy" is an involved task in indirect taxes. A simple approach is to levy and collect the tax at a single point

© Springer Nature Switzerland AG 2019
K. Chen et al. (Eds.): BigData 2019, LNCS 11514, pp. 95–109, 2019.
https://doi.org/10.1007/978-3-030-23551-2_7

in the value chain, for example, the point of final consumption. The retail sales tax (RST) in the United States of America is an example. Single point of the levy is easy for administering but it has a few flaws. Many developing countries have a high concentration of informal economic activities at the consumption points. For example, the market share of informal economic activities in India is almost 50%. In these countries, there is a major risk of loosing out tax at the final consumption point. Sales tax can be sidestepped by taking the goods out of the value chain right at the onset. This will result in the creation of a parallel economy by keeping a major part of the value chain outside the regulatory authority's watch. One approach towards handling this problem is by following a multipoint taxation system, such as the Value-added Tax (VAT) and the Goods and Services Tax (GST) [5]. Goods and Services Tax, which is implemented in India from July 2017, is a comprehensive, multi-stage, destination-based tax that is levied on every value addition. This tax has replaced many indirect taxes that were previously existed in India.

1.1   Multipoint Taxation System (VAT and GST)

In this system, the tax is levied incrementally in each stage of the production depending upon the value added to goods in the corresponding production phase. Tax is levied at each phase of the production, such that tax paid on purchases(input tax) will be given as set-off for that tax levied on the sales (output tax) [8]. Figure 1 shows how the tax is collected incrementally in this system.

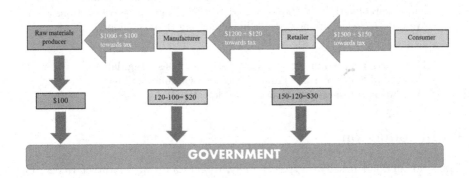

Fig. 1. Multipoint taxation system

- In this example, the manufacturer purchases some raw material of value 1000$ from the raw material dealer, by paying 100$ as tax at 10% tax-rate. The raw materials dealer remits to the government the tax amount that he has collected.
- Then the retailer purchases the processed goods from the manufacturer for, say, 1200$. An amount of 120$ is then paid to the manufacturer as a tax. The manufacturer pays the government the difference between the tax he had collected

(from the retailer) and the tax he has paid (to the raw materials producer) (120\$ − 100\$ = 20\$).

- The consumer then buys the finished goods from the retailer for 1500\$ by paying a tax of 150\$. By following the same argument as given in the previous steps, the retailer pays 30\$(i.e., 150\$ − 120\$) to the government.

It can be easily calculated that the total tax received by the government is 150\$, and it is indirectly paid completely by the consumer of goods. Hence, raw material dealer, manufacturer, and retailers are representatives of the government to collect the tax.

This method ensures market-driven checks and balances or compliance regime which is difficult to achieve. At each node in the value chain, the purchaser and seller duo would have contradicting goals towards their tax liabilities. The seller tries to understate his sales while the purchaser tries to overstate purchases. This contradicting approach ensures market-driven checks and balances.

1.2 Tax Evasion Methods in Multipoint Taxation System

Circular Trading in GST and VAT: In GST and VAT, the market-driven checks and balances did not work in an expected manner. In the majority of tax evasion cases, business dealers, in their monthly tax-returns, deliberately manipulate their actual business transactions motivated by the amount of profit gained by evading tax. Invoice trading is a method to evade tax [2], where a dealer sells their goods to the end user without issuing the invoice but collecting the tax. Later, he/she issues a fake invoice to a third party, who uses it towards increasing their input tax credit. This will minimize the amount of tax they have to pay in the form of cash(the difference between the tax they collected at the time of sales and the tax they paid at the time of purchases) to the government. These tax manipulations can be spotted by the tax enforcement officials. To hide these manipulations, malicious dealers create a well-entrenched "racket," where a large number of bogus firms(shell firms) are created to manipulate the title of goods in the first place and then follow it up by making fake transactions among them to outwit easy systemic detection. Malicious dealers, show huge fake sales and purchases among malicious dealers and dummy dealers(shell firms) without any significant *value-added* as given in Fig. 2.

In Fig. 2, illegitimate transactions corresponding to fake invoices (invoice trading) are shown using red-lines. These are from dealers, x to q, x to z and q to z. Dealers q and z use these fake invoices towards minimizing their tax liability. With the motivation of confusing the tax enforcement officers, these dealers superimpose several fake transactions(dummy transactions) on these illegitimate transactions, which are shown using gray lines. Note that these dealers superimpose fake transactions such that the tax liability of any of the dealer due to the fake transactions is zero, *i.e.*, an amount of tax paid on the fake purchases is equal to the amount of tax collected on the fake sales.

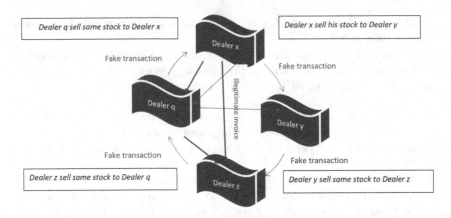

Fig. 2. Circular flow of sales/purchases

Since the value-addition due to the fake transactions is equal to zero, they do not pay any tax on these fake transactions, but, rather they create confusion to the tax officials about the illegitimate transactions. It is important to note that there is a huge amount of fake sales and purchases transactions among malicious and dummy dealers when compared to genuine sales and purchases transactions with the others. This type of technique used to evade tax is known as *circular trading* [9,10,21]. Hence the malicious dealers complicate the process of detecting their illegitimate transactions (invoice trading).

Carousel Fraud: Carousel fraud is a method of stealing public money by exploiting the VAT-free trade arrangements between European Union member countries. An organized crime groups will import goods from another country, then sell them by charging VAT to the customer but absconding with VAT instead of passing it to the government. To make this process undetectable, these groups buy and sell the goods multiple number of times between bogus companies before the final transaction where the VAT is stolen [4,18].

Carousel fraud and circular trading have a lot of characteristics in common. The solutions for circular trading can be extended to carousel fraud. In this paper, we work on circular trading.

1.3   Motivation for This Work

Manually, it is impractical for the tax officials to detect illegitimate transactions in circular trading due to the enormous size of the tax department's database, complicated sequences of sales and purchases transactions by the malicious dealers, the unknown identity of the traders doing these manipulations, *etc.* These challenges call for sophisticated big data and graph theoretic techniques. We used the *RHadoop* framework [6] for implementation.

The following gives a brief account of the paper structure. In Sect. 2, we describe several existing approaches that are used to perform cluster analysis on problems similar to that of ours. In Sect. 3, the problem is formulated as detecting communities in social networks and removing cycles created by fake transactions. In Sect. 4, we outline the experimental setup and results obtained from this work. We implemented these algorithms for the Commercial Taxes Department, Government of Telangana, India.

2   Related Work

Circular trading is a notorious problem in stock markets. In [21], Palshikar et al. proposed a highly customized algorithm for identifying colluding sets in stock trading. In [27], Wang, et al. proposed an algorithm to identify colluding sets in the instrument of future markets. In [13], Islam, et al. had given an algorithm for identifying collusion sets and cross trading collusion sets.

In [20], Nigrini et al. suggested statistical methods which can be used in the initial stages of the auditing. These techniques are based on Benford's Law, a unique characteristic of tabulated numbers. This law gives the expected probability of the digits in tabulated data. In [1], Arben Asllani et al. proposed a method that can be used by charted accountants to detect accounting fraud.

In [16], Klymko et al. have given an undirected edge weighting method based on directed triangles to detect communities in directed networks. They proposed a new measure on the quality of the communities in social networks depending on the number of 3-cycles that are span across communities. They showed that the resulting communities have fewer 3-cycles cuts. In [14,23], the author showed the significance of triangles in community detection in an undirected networks. In [15], Khadivi et al. showed that proper assignment of weights to the edges of a social network could improve community detection. They used this weighting as an initial step for the Newman greedy modularity optimization algorithm. In [17], the authors have proposed a method which can identify classification rules to detect fraudulent samples. They discovered spatial relationships of fraud and non-fraud financial statements. In [12], the authors have proposed a clustering based data mining algorithm to find outliers in taxation data. In [11], the authors have used clustering algorithms to identify a group of taxpayers, and then they have used several classification models to detect a potential user of false invoices in a given year.

In [6], Dean and Ghemawat explained MapReduce programming model for processing large data sets. In [3], Behera et al. had explained the implementation of random walk based graph clustering algorithm using Map-Reduce framework. In [25], Rajaraman et al. had given algorithms to handle massive data sets.

3   Problem Statement and Solution

It is impractical for the tax officials to detect illegitimate transactions in circular trading manually. Our objective is to design an algorithm to detect illegitimate transactions and the set of a dealer doing these transactions. We follow the below four-step approach to solve the problem.

- **Step 1**: Construct an edge weighted directed graph from the way bill data base, where vertices correspond to dealers, and weights of directed edges are defined by the number of fake transactions, which are identified by Benford's analysis.
- **Step 2**: Convert this edge weighted directed graph into an edge weighted undirected graph.
- **Step 3**: Identify the groups of a dealer who perform excessive trade among themselves, as compared to the sales and purchases with other dealers. The problem is formulated as finding fraudulent communities in a social network.
- **Step 4**: Remove cycles formed by fake transactions within each group of these dealers.

3.1   Step 1: Construction of Sales Transaction Graphs

Waybill Database: Table 1 is a sample of a waybill data base. Each row corresponds to a sales transaction. Each row contains seller name, purchaser name, time of sales and value of sales.

Table 1. Waybill database

S.no	Seller	Purchaser	Time	Value
1	Tax Payer X	Tax Payer Y	2019/01/04/15:20	13000
2	Tax Payer Z	Tax Payer U	2019/01/04/17:00	19000
3	Tax Payer X	Tax Payer U	2019/01/05/19:00	15000
4	Tax Payer Y	Tax Payer Z	2019/01/05/17:00	15000
5	Tax Payer Z	Tax Payer X	2019/01/05/15:30	13000

The actual database contains many more details like type of goods, the rate of tax, the quantity of goods, vehicle used for transporting the goods, vehicle number, transporter name, invoice number, UOM (unit-of-measure), inserted date, etc. The data we had taken contains several million rows.

Benford Analysis: Benford's law, which is also known as the first digit law, is a statistical technique for fraud detection [1,7,20]. This law intrigued mathematicians for over a century. This law gives the probability of the leading digit in a naturally occurring numeral data.

The Benford's law states that for any numerical data with a distribution of numbers spanning several orders of magnitude(an order of magnitude is an approximate measure of the number of digits that a number has in the commonly-used base-ten number system), the probability of a number starting with the digit d is given by $log_{10}(1 + 1/d)$, where $d \in \{1, 2, ..., 9\}$.

Mean absolute deviation (MAD) is a statistical method which can be used to find whether the data's first digits follow the probability distribution given by Benford's law. Mean absolute deviation $MAD = \sum_{j=1}^{m}(OP_j - EP_j)/m$, where OP_j is the observed probability of jth bin, EP_j is the expected probability of jth bin, and m is the total number of bins (in this case it is equal to 9). Based on the MAD value, we can find the conformity between expected probability and observed probability as given below [19].

- MAD value between 0.000 to 0.004 says "Close conformity"
- MAD between 0.004 to 0.008 says "Acceptable conformity"
- MAD between 0.008 to 0.012 says "Marginally acceptable conformity"
- MAD greater than 0.012 says "Nonconformity"

Sales Transaction Graph: We use waybill database to construct an edge weighted directed social network denoted by $G_d = (V_d, E_d)$, where V_d is the vertex set (each dealer corresponds to a vertex), and E_d denotes the set of weighted directed edges. We name this social network as *sales transaction graph*. Below we propose a method to assign weights to the edges.

Let m be the number of sales transactions in the waybill database from dealer vertex x to dealer vertex y and $v_1, v_2, v_3, \ldots, v_m$ be values of these sales. Let $\beta(xy)$ be the MAD value of the first digit Benford's analysis on $v_1, v_2, v_3, \ldots, v_m$. Based on the value of $\beta(xy)$, we can establish the conformity between expected and observed distribution.

The weight $w(xy)$ of the edge from vertex x to vertex y in graph G_d is given by $w(xy) = (m * \sum_{i=1}^{m} v_i)/(m + \sum_{i=1}^{m} v_i) * e^{1000*\beta(xy)}$. Note that lesser edge weights are assigned for the edges with less number of transactions or less sum of the values of the transactions [24]. The weight of the edge xy increases exponentially with $\beta(xy)$, i.e., more weight is assigned for a lesser conformity between expected distribution and observed distribution.

3.2   Step 2: Construction of Weighted Undirected Graph

Majority of work in community detection has been done on undirected graphs. In this paper, we propose a method to convert an edge weighted directed graph into an edge weighted undirected graph.

There are several metrics to measure the quality of a community. One major idea is that flow tends to stay within the community. Hence, cycles in a graph play an important role in community detection. In detecting communities in circular trading, 2-cycles and 3-cycles play an important role. We propose a weighting scheme to turn an edge weighted directed graph to an edge weighted undirected graph. The weight given for an edge is based on triangles and two cycles in which this edge is involved [15, 16, 23].

In the following, we will explain how to construct an edge weighted undirected graph $G_u = (V_u, E_u)$ from an edge weighted directed graph G_d described in Subsect. 3.1. Let $C = (a, b, c)$ be a cycle in G_d. Cycle C can be any one of the four types of cycles shown in Fig. 3. The weight of cycle C is defined as follows.

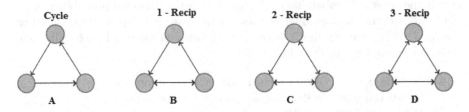

Fig. 3. Different types of 3-Cycles

- If it is of *Type a*, then the weight of the cycle C is given by $W(C) = min\{w(ab), w(bc), w(ca)\} * 1$
- If it is of *Type b*, then the weight of the cycle C is given by $W(C) = min\{w(ab), w(bc), w(ca)\} * 1.5$
- If it is of *Type c*, then the weight of the cycle C is given by $W(C) = min\{w(ab), w(bc), w(ca)\} * 1.75$
- If it is of *Type d*, then the weight of the cycle C is given by $W(C) = min\{w(ab), w(bc), w(ca)\} * 1.875$

The weights 1, 1.5, 1.75 and 1.875 are given to 3-cycles of types a, b, c, d respectively. These are chosen empirically based on the clustering performance. The number of reciprocal edges in a triangle conveys the strength of circular trading. Hence, we gave more weight to a 3-cycle with more number of reciprocal edges.

Suppose directed edge ab is in cycles $C_1, C_2, \ldots, C_m$ and directed edge ba is in cycles $D_1, D_2, \ldots, D_n$. Then the weight of an undirected edge ab in graph G_u is given by the maximum among the following three values

- $max\{W(C_1), W(C_2), \ldots, W(C_m)\}$
- $max\{W(D_1), W(D_2), \ldots, W(D_n)\}$
- $min\{w(ab), w(ba)\} * 2$

3.3   Step 3: Community Detection

We use WalkTrap algorithm to detect communities. WalkTrap algorithm is a hierarchical agglomerate clustering algorithm and it uses a distance measure based on random walks [22]. It is based on the assumption that a random walker would spend a longer time inside a strong community due to the high density of edges within the community. This algorithm measures the similarity between vertices and between communities by defining a distance between them. This distance measure is calculated from the probabilities that the random walker moves from one vertex to another in a fixed number of steps.

Distance Between Communities. Let us consider random walks of a given length t on graph G. Let p_{ij}^t be the probability of reaching vertex j from vertex i in a random walk of length t. Value of t should be large enough to capture the community structure of G but not too large to reach a stationary distribution. Generally, the value of t is between three and six. The basic idea behind this algorithm is two vertices of the same community tend to *see* all the other vertices in the same way. Thus if vertices i and j are in the same community, we can expect that $\forall k$, $p^t ik \cong p^t jk$. Then the distance between vertices i and j can be defined as $\sqrt{(\sum_{k=1}^{n} \frac{(p_{ik}^t - p_{jk}^t)^2}{d(k)})}$, where $d(k)$ is the degree of vertex k [22]. One can generalize the distance between vertices to a distance between communities in a straightforward way.

3.4   Step 4: Removing Cycles in Each Cluster

Consider any community (cluster) C given by the community detection algorithm. Using the waybill database explained earlier, we construct a directed edge-labeled multi-graph called *sales and purchase graph*, denoted by $G_{sp} = (U, E, \gamma)$, where U is the set of vertices (each vertex corresponds to a dealer in C), E is the set of labeled directed edges (an edge from vertex x to vertex y corresponds to a sales transaction from x to y) and γ is the function that associates a 2-tuple for each labeled edge, where the first element of the tuple is the time of sales of this transaction and the second element is the value of sales of this transaction.

Following are few notations we use. Note that each edge in the graph has two parameters, one is the time of sales and the other is the value of sales. The *end_time* of a cycle is defined as the time of most recent sales transaction among all the sales transactions corresponding to the edges in the given cycle. The *start_time* of a cycle is defined as the time of least recent transaction among all the sales transactions corresponding to the edges in the given cycle. The *time_gap* of a cycle is defined as the difference between *end_time* and *start_time*. The *maxval* of a cycle is defined as the maximum value among values of all the transactions corresponding to the edges in the given cycle. The *minval* of a cycle is defined as the minimum value among values of the transactions corresponding to the edges in the given cycle. The *valgap* of a cycle is defined as the difference between *maxval* and *minval*. Let the *trust score* of a cycle is defined as *time_gap* * *valgap*.

From in-depth research by taxation authorities, it is observed that *time_gap* and *valgap* of any fake sales cycle are very small, which means the *trust score* of any fake cycle is small. Our motive is to remove all fake cycles from the *sales and purchase graph*. Then the remaining graph will be a directed acyclic graph (DAG). Note that the resultant directed acyclic graph contains all suspicious transactions. This makes fraud detection process simpler which allows us to do a deeper analysis on suspicious transactions to identify tax evaders. Below we give a brief sketch of the fake cycles removal algorithm.

1. Select a cycle D in sales and purchase graph G_{sp} with the following conditions:
   - *Condition 1: end_time* of D is minimum among all the cycles in G_{sp}
   - *Condition 2:* With respect to the condition one, *trust score* of D is minimum
   - *Condition 3:* With respect to the condition two, *length* of D is minimum
2. Let y be the minimum of the values of sales of all the edges in D. Subtract y from values of sales of all edges in D.
3. Remove any edge from D whose value of sales becomes zero.
4. Repeat steps one to three, as long as G_{sp} contains a cycle.

3.5   Algorithms

Detecting and Managing Outliers: According to Benford's analysis, the probability of nine occurring as the first digit is 0.046 [19]. We need at least twenty-two transactions between any pair of dealers to get a valid Benford's score. As part of data cleansing, we remove sales transactions between pairs of dealers (vertices) if the number of sales transactions between them is less than twenty-two. If the value of any sales transaction in waybill database is more than *third quantile plus 1.5 times the inter-quantile range of values of sales transactions*, then replace the value of this sales transactions by *third quantile plus 1.5 times the inter-quantile range of values of sales transactions* [2].

Algorithms: Algorithm 1 is a community detection algorithm. First, we apply this algorithm to detect communities. Removing outliers and construction of directed graph are highly time consuming operations in this algorithm due to millions of purchase and sales transactions. These operations are parallelized. We used Map-Reduce framework to implement these operations. Later we apply Algorithm 2 to identify illegitimate transactions. Note that both algorithms are polynomial time algorithms.

4   Case Study

4.1   Experimental Setup

We used the R programming language for data mining and Hadoop framework for storing data. We used the *RHadoop* open source analytics solution to integrate R programming language with *Hadoop*.

Data: WayBill Data
Result: Set of Communities

Perform outlier cleansing;
`# This is explained in 3.5.;`

Construct a directed graph G_d;
`# This is explained in 3.1. Note that Benford's analysis has to be`
`performed on the values of sales transaction before outlier`
`cleansing.;`

Construct an undirected graph G_u;
`# This is explained in 3.2;`

Find communities in G_u using WalkTrap algorithm;
`# If any community is bigger than eight vertices, perform sub-`
`community detection on this community;`
`}`

Algorithm 1. Community detection algorithm

4.2   Identifying Communities

In our data set, there are 0.6 million dealers. Size of our data set is 1.5 TB. Figure 4, shows the business among some of these dealers. We applied the Algorithm 1 on this data set and obtained several communities, which are doing heavy circular trade. Figure 5, shows a few communities obtained. We used two measures namely modularity and coverage to validate the clustering. Modularity and coverage are 0.74 and 0.82 respectively.

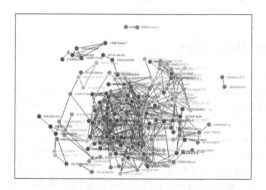

Fig. 4. Complex network of sales and purchases

Data: sales and purchase graph G_s

Result: Forest G_t, which is obtained by removing all cycles in G_s

G_t = Edgeless graph whose vertex set is $V(G_s)$;

Let $l_1, l_2, \ldots, l_m$ be a sequence of all edges in G_s ordered by non decreasing order of time of sales;

for $i = 1 \ldots m$ **do**
    insert the edge l_i in the graph G_t;

    while *(G_t contains cycle)* **do**
        Assume that the edge l_i is from vertex b to vertex a in G_t;

        Let $P_1, P_2, \ldots, P_k$ be set of the path from a to b in G_t;

        Let sp_i, vg_i be *time_gap* and *valgap* of cycle C_i formed by path P_i along with the edge ba, for $1 \leq i \leq k$;

        Let $spdiff = \max\{sp_1, sp_2, \ldots, sp_k\}$-$\min\{sp_1, sp_2, \ldots, sp_k\}$;

        Let $valgapdiff = \max\{vg_1, vg_2, \ldots, vg_k\}$-$\min\{vg_1, vg_2, \ldots, vg_k\}$;

        Let *normalised trust score* of cycle C_i be $(sp_i/spdiff) * (vg_i/valgapdiff)$ for $1 \leq i \leq k$;

        Let C_j be a cycle, where $1 \leq j \leq k$, such that *normalized trust score* is minimum;
        `# This cycle can be identified in polynomial time;`

        Let p be the minimum among the price of sales of all edges in C_j;

        Subtract p from the price of sales of all edges in C_j;

        Remove all edge from G_t whose price of sales is zero;

    end
end

Algorithm 2. Cycle removal algorithm

4.3   Identifying Illegitimate Transactions

We had taken one community with four dealers which is shown in Fig. 6. These four dealers are doing heavy circular trade among themselves. Their sales, purchase and tax details are shown in Fig. 7. Total tax paid by these four dealers is Indian rupees 0.03 million which are shown in column seven. The tax they collected on sales(output tax) is Indian rupees 367 million as shown in column six. They set-off this entire tax collected with the tax they paid on purchases(input tax) which is shown in column four. In genuine Iron and Steel, the business ratio between the input tax and output tax will be less than 0.95, but here it is almost one. We applied Algorithm 2 on this community to remove fake cycles and identify illegitimate transactions. When the tax authorities physically visited the premises of these companies, they identified that these are shell companies.

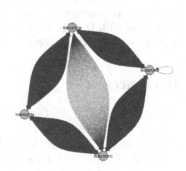

Fig. 5. Experimental result   **Fig. 6.** Cluster of four dealers

S No	Dealer	Purcahse Amount	Input Tax Credit	Sales Amount	Output Tax	Tax Payment
1	2	3	4	5	6	7
1	50126652656	1760.68	88.03	1731.39	86.57	0.00
2	50410281020	1998.65	100.23	2021.42	101.21	0.00
3	50484623784	1711.67	85.58	1712.72	85.64	0.03
4	50561583571	1996.01	99.80	1902.17	95.11	0.00
	TOTAL	7467.01	373.64	7367.7	368.53	0.03

Fig. 7. Business details

5   Conclusion

Here we studied a widely practiced tax evasion method in GST called *circular trading*. *Circular trading* is a tax evasion practice where a set of malicious dealers do heavy fake sales and purchase transactions among themselves that go around in a circular manner in a very short time-duration without any meaningful *value-addition*. They practice this technique to hide illegitimate transactions. We addressed the problem of identifying the cluster of dealers who do excessive fake trade among themselves and illegitimate transactions performed by them. We implemented this technique using *RHadoop* big data framework for the Commercial Taxes Department, Government of Telangana, India. Our results are helping the tax authorities to effortlessly identify illegitimate transactions and take legal action against those who are doing these transactions. As future work, we plan to work on developing sophisticated algorithms that detect colluding communities by exploiting the different patterns made by the fraudulent dealers.

Acknowledgment. We would like to express our deep thanks towards the government of Telangana, India, for allowing us to use the Commercial Taxes Data set and giving us constant encouragement and financial support. This work has been supported by Visvesvaraya Ph.D. Scheme for Electronics and IT, Media Lab Asia, grant number EE/2015-16/023/MLB/MZAK/0176.

References

1. Arben Asllani, M.N.: Using Benford's law for fraud detection in accounting practices. J. Soc. Sci. Stud. **1**, 129–143 (2014)
2. Baesens, B., Vlasselaer, V., Verbeke, W. (eds.): Fraud Analytics Using Descriptive, Predictive, and Social Network Techniques: A Guide to Data Science for Fraud Detection. Wiley, Hoboken (2015). ISBN 978-1-119-13312-4
3. Behera, R., Rath, S., Misra, S., Damasevicius, R., Maskeliunas, R.: Large scale community detection using a small world model. Appl. Sci. **7**, 1173 (2017)
4. Borselli, F., Fedeli, S., Giurato, L.: Digital VAT carousel fraud: a new boundary for criminality. Tax Notes International (2015)
5. Dani, S.: A research paper on an impact of goods and service tax (GST) on indian economy. Bus. Econ. J. **7**, 264 (2016). ISSN 2151–6219
6. Dean, J., Ghemawat, S.: MapReduce: simplified data processing on large clusters. In: Proceedings of the 6th Conference on Symposium on Opearting Systems Design & Implementation, OSDI 2004, vol. 6, p. 10. USENIX Association, Berkeley (2004). http://dl.acm.org/citation.cfm?id=1251254.1251264
7. Durtschi, C., Hillison, W., Pacini, C.: The effective use of Benford's law to assist in detecting fraud in accounting data. J. Forensic Account. **V**, 17–34 (2004)
8. Dutta, R., Kumar, B.: Value added tax scams and introduction of the goods and services tax. Econ. Polit. Wkly. **53**(44) (2018)
9. Franke, M., Hoser, B., Schröder, J.: On the analysis of irregular stock market trading behavior. In: Preisach, C., Burkhardt, H., Schmidt-Thieme, L., Decker, R. (eds.) Data Analysis, Machine Learning and Applications, pp. 355–362. Springer, Heidelberg (2007). https://doi.org/10.1007/978-3-540-78246-9_42. ISBN 978-3-540-78239-1
10. Golmohammadi, K., Zaiane, O., Díaz, D.: Detecting stock market manipulation using supervised learning algorithms. In: Data Science and Advanced Analytics, pp. 435–441. IEEE, November 2014. http://ieeexplore.ieee.org/document/7058109/, ISBN 978-1-4799-6991-3
11. González, P.C., Velásquez, J.D.: Characterization and detection of taxpayers with false invoices using data mining techniques. Expert Syst. Appl. **40**(5), 1427–1436 (2013)
12. Huang, S.Y., Tsaih, R.H., Yu, F.: Topological pattern discovery and feature extraction for fraudulent financial reporting. Expert Syst. Appl. **41**(9), 4360–4372 (2014)
13. Islam, N., Rafizul Haque, S., Masudul Alam, K., Tarikuzzaman, M.: An approach to improve collusion set detection using MCL algorithm. In: Computers and Information Technology, pp. 237–242. IEEE, December 2009. http://ieeexplore.ieee.org/abstract/document/5407133/, ISBN 978-1-4244-6284-1
14. Berry, J.W., Hendrickson, B., LaViolette, R.A., Phillips, C.A.: Tolerating the community detection resolution limit with edge weighting. Phys. Rev. E **83**, 056119 (2011)

15. Khadivi, A., Ajdari Rad, A., Hasler, M.: Network community-detection enhancement by proper weighting. Phys. Rev. E **83**, 046104 (2011). https://doi.org/10.1103/PhysRevE.83.046104
16. Klymko, C., Gleich, D.F., Kolda, T.G.: Using triangles to improve community detection in directed networks. ASE BIGDATA/SOCIALCOM/CYBERSECURITY Conference, Stanford University abs/1404.5874 (2014)
17. Liu, B., Xu, G., Xu, Q., Zhang, N.: Outlier detection data mining of tax based on cluster. Phys. Procedia **33**(44), 1689–1694 (2012)
18. Frunza, M.-C.: Aftermath of the VAT fraud on carbon emissions markets. J. Financ. Crime **20** (2013)
19. Mark Nigrini, J.T.W. (ed.): Benford's Law: Applications for Forensic Accounting, Auditing, and Fraud Detection. Wiley, Hoboken (2012). ISBN 978-1-118-15285-0
20. Nigrini, M.J., Mittermaier, L.J.: The use of Benford's law as an aid in analytical procedures. Audit.: J. Pract. Theory **41**, 52 (1997)
21. Palshikar, G., Apte, M.: Collusion set detection using graph clustering. Data Min. Knowl. Discov. **16**, 135–164 (2008). https://doi.org/10.1007/s10618-007-0076-8. ISSN 1384-5810
22. Pons, P., Latapy, M.: Computing communities in large networks using random walks. In: Yolum, I., Güngör, T., Gürgen, F., Özturan, C. (eds.) ISCIS 2005. LNCS, vol. 3733, pp. 284–293. Springer, Heidelberg (2005). https://doi.org/10.1007/11569596_31
23. Prat-Pérez, A., Dominguez-Sal, D., Brunat, J.M., Larriba-Pey, J.L.: Shaping communities out of triangles. In: ACM International Conference Proceeding Series, July 2012
24. Jarvis, R.A., Patrick, E.A.: Clustering using a similarity measure based on shared nearest neighbors. IEEE Trans. Comput. **C-22**(11), 1025–1034 (1973)
25. Rajaraman, A., Ullman, J.D.: Mining of Massive Datasets. Cambridge University Press, New York (2011)
26. Schenk, A., Oldman, O. (eds.): Value Added Tax: A Comparative Approach. Cambridge University Press, Cambridge (2007). ISBN 978-1107617629
27. Wang, J., Zhou, S., Guan, J.: Detecting potential collusive cliques in futures markets based on trading behaviors from real data. Neurocomputing **92**, 44–53 (2012)

Multi-step Short Term Traffic Flow Forecasting Using Temporal and Spatial Data

Hao Peng$^{(\boxtimes)}$ and John A. Miller

Department of Computer Science, University of Georgia, Athens, GA, USA
penghga@uga.edu, jam@cs.uga.edu

Abstract. Short term traffic flow forecasting is valuable to both governments for designing intelligent transportation systems and everyday commuters or travelers who are interested in the best routes to their destinations. This work focuses on forecasting traffic flow in major freeways in southern California using large amounts of data collected from the Caltrans Performance Measurement System. Both statistical models and machine learning models are considered. The statistical models include seasonal ARIMA, seasonal VARMA, exponential smoothing and various regression models. The machine learning models include Support Vector Regression, feed forward Neural Networks, and Long Short-Term Memory Neural Networks. Forecasting is performed in both a univariate manner by relying on the historical temporal data of a particular sensor as well as in a multivariate manner by considering a neighborhood of three closely located sensors. Multivariate forecasters generally improve upon their univariate counterparts.

Keywords: Traffic flow forecasting · Big data analytics ·
Time series analysis · Seasonal ARIMA · Seasonal VARMA ·
Exponential smoothing · Regression · Support Vector Regression ·
Neural Networks · LSTM

1 Introduction

Traffic flow forecasting is an important aspect in designing intelligent transportation systems for cities and highways. It is also of great interests to everyday travelers who may desire to know in advance the congestion levels of roads and the amount of time it would take to reach their destinations. Much research has been devoted into studying this topic in recent years, as can be evident in very recent work such as [12,20,27,29]. The fruits of such studies can be of good use to city planners in governments, traffic app developers as well as everyday commuters and travelers.

The dawn of the big data era has also greatly facilitated the advancements in research on traffic forecasting. Seeing the need to collect large amounts of high quality and high resolution traffic data, numerous states in the United

© Springer Nature Switzerland AG 2019
K. Chen et al. (Eds.): BigData 2019, LNCS 11514, pp. 110–124, 2019.
https://doi.org/10.1007/978-3-030-23551-2_8

States have invested in deploying great number of traffic sensors on their busiest roads and highways. The Caltrans Performance Measurement System (PeMS)[1] from the state of California is an example of such systems. High resolution traffic data such as flow and speed are collected in real-time from more than 39000 sensors deployed in major urban areas and highways across the state. This work is devoted to studying traffic flow forecasting using PeMS data collected in southern California during the entire year of 2018.

The vast majority of existing literature on the topic of traffic forecasting has devoted to forecasting in the immediate short terms, such as a couple of minutes ahead. It is certainly justified, as the immediate short terms can usually best capture the dynamic nature of traffic situations and are usually of the greatest interests. For example, a commuter would be very interested in the optimal routes to avoid the most traffic congestions during morning rush hours; or the amount of time, hopefully in minutes, for the commuter to arrive at his or her workplace. Longer term traffic forecasts can certainly be done by relying more on the historical data of a particular location but may suffer from relatively poorer accuracies due to larger time gaps (e.g., forecasting 24 h in advance may need to heavily rely on historical averages, but if exceptional circumstances occur, such as an accident or rainy weather, then clearly the forecasts produced 24 h ago may not be as reliable).

Many researchers have exclusively studied on 1-step ahead forecast, such as in [7,14,15,17]; in other words, if the data are of 15 min resolutions, then forecasts are produced for only 15 min ahead. Others, especially recently, have also studied multi-step traffic forecasting, from minutes to a couple of hours ahead, such as in [12,16,29].

Univariate forecasting, meaning producing forecasts by relying on historical data from one particular sensor alone, is also most prolific in the literature, such as in [10,12,15,17,26]. Multivariate forecasting usually involves using data from multiple spatially dependent sensors to produce improved forecasts than the univariate counterparts, such as in [3,16,29]. Very recently, some researchers have also chosen to simply give data from very large numbers of sensors to a deep Neural Network and task it to determine and establish any dependencies among the data [16,29]. Some studies have also included external variables such as weather data into their forecasting models, such as in [12,14].

This work can be thought of as an extension to a very recent previous work in [21]. Additional popular forecasting models are included and a new multivariate forecasting experiment is conducted. More details on the improvements upon our previous work are included in the Related Work section. The contributions of this work are as follows: (1) to evaluate the effectiveness of commonly used statistical and machine learning models on univariate traffic flow forecasting using large amounts of temporal data; (2) to study the impacts of incorporating spatially dependent data into multivariate forecasting models; (3) to examine the performance of multi-step forecasts in the short term, which is generally very dynamic and volatile. (4) to provide a reference for the relative performance of popular traffic flow forecasting models in both univariate and multivariate settings.

---

[1] http://pems.dot.ca.gov/.

Most forecasting models used in this work are provided by the SCALATION project [19]. It is a open source, MIT licensed, Scala-based project designed for analytics and simulation using big data. For more details, please visit http://www.cs.uga.edu/~jam/scalation.html. The only exceptions are the Neural Networks models, which are provided by Keras [5] using the Tensorflow [1] backend.

The rest of this paper is organized as follows: Sect. 2 discusses the basic background on various statistical and machine learning models included in this study. Section 3 is about Related Work in traffic forecasting. Section 4 explains the detailed experimental setup and performance evaluations. Finally, Sect. 5 concludes the paper and offers potential directions for future work.

2   Background

In general, a forecasting model may take on the form of

$$y_t = f(X, B) + \epsilon_t, \tag{1}$$

where y_t the response of interest at time t (e.g., the traffic flow at 8:00AM); X is the set of inputs (e.g., traffic flow data in the recent past); B is the set of parameters; f is a function that maps X and B to a forecasted value at time t, often denoted as $\hat{y}_t$; and ϵ_t is the residual at time t.

Though the exact form of the function f, the set of parameters B, and the set of inputs X can differ greatly for various forecasting models, their common goal is to produce forecasted values that are as close to the actual values as possible across multiple time instances and minimize some type of norm of the residuals such as the Sum of Squared Errors (SSE).

The model in Eq. 1 may be generalized into the multivariate case consisting of m time series as follows:

$$\mathbf{y}_t = g(X, B) + \boldsymbol{\epsilon}_t, \tag{2}$$

where the response $\mathbf{y}_t$ and the residuals $\boldsymbol{\epsilon}_t$ have all been generalized into dimension m. The function $g(X, B)$ now maps X and B to a vector of forecasted values, one for each of the m time series.

2.1   Statistical Models

Statistical models generally involve formalization of equations that try to explain the relationships among various variables based on certain assumptions. Commonly use statistical forecasting models include seasonal Autoregressive Integrated Moving Average (ARIMA) model [2]; its multivariate generalization seasonal Vector Autoregressive Moving Average (VARMA) model [24]; exponential smoothing model [9,28]; as well as regression models.

Seasonal ARIMA. In reference to Eq. 1, the seasonal Autoregressive Moving Average model expresses $f(X, B)$ as

$$f(X, B) = c + \sum_{i=1}^{p} \phi_i y_{t-i} + \sum_{i=1}^{q} \theta_i \epsilon_{t-i} + \sum_{i=1}^{P} \Phi_i y_{t-il} + \sum_{i=1}^{Q} \Theta_i \epsilon_{t-il}, \qquad (3)$$

where the set of inputs X includes the i-th lagged values of the response y_{t-i} and residual ϵ_{t-i} as well as the seasonal lagged values y_{t-il} and ϵ_{t-il} of seasonal period l; the set of parameters B contains an intercept c, p autoregressive parameters ϕ's, q moving average parameters θ's, and their seasonal counterparts, P Φ's and Q Θ's. Differencing of order d or seasonal differencing of order D may also be applied to the time series to stabilize the mean before fitting the parameters. Notation wise, it is common to express a seasonal ARIMA model as SARIMA $(p, d, q) \times (P, D, Q)_l$.

Seasonal VARMA. The seasonal Vector Autoregressive Moving Average model is the multivariate generalization of the seasonal ARIMA model. Instead of only relying on the lagged values of one time series to make forecasts, the seasonal VARMA model incorporates lagged values from m time series to help make forecasts for each time series. The seasonal VARMA model expresses $g(X, B)$ in Eq. 2 as

$$g(X, B) = \mathbf{c} + \sum_{i=1}^{p} A_i \mathbf{y}_{t-i} + \sum_{i=1}^{q} M_i \boldsymbol{\epsilon}_{t-i} + \sum_{i=1}^{P} U_i \mathbf{y}_{t-il} + \sum_{i=1}^{Q} O_i \boldsymbol{\epsilon}_{t-il}, \qquad (4)$$

where $\mathbf{c}$ is a vector of dimension m representing intercepts for the m time series in the model; the parameter matrices A's, M's, U's, and O's are all of dimensions $m \times m$ and are the multivariate generalizations of ϕ's, θ's, Φ's, and Θ's in Eq. 3, respectively; Similarly, $\mathbf{y}$ and $\boldsymbol{\epsilon}$ are both vectors of size m, representing the values and residuals of all m time series.

Exponential Smoothing. Another univariate forecasting model is exponential smoothing, for which the $f(X, B)$ in Eq. 1, when given data up to time $t - 1$, may be calculated as

$$f(X, B) = s_{t-1} + d_{t-1} + a_{t-l}, \qquad (5)$$

where the smoothed value s, the trend factor d, and the additive seasonal factor a may be recursively computed as

$$s_{t_i} = \alpha(y_{t_i} - a_{t_i-l}) + (1 - \alpha)(s_{t_i-1} + d_{t_i-1}),$$
$$d_{t_i} = \beta(s_{t_i} - s_{t_i-1}) + (1 - \beta)d_{t_i-1}, \qquad (6)$$
$$a_{t_i} = \gamma(y_{t_i} - s_{t_i}) + (1 - \gamma)a_{t_i-l},$$

where $t_i \in [1, t)$; the set of inputs X includes all values of the time series up to time $t - 1$; the set of parameters B contains α, β and γ, known as smoothing parameters bounded between 0 and 1.

Regression. Regression models such as linear regression and polynomial regression are usually very efficient and effective. The linear regression model may express $f(X, B)$ in Eq. 1 as

$$f(X, B) = \mathbf{b} \cdot \mathbf{x}_t, \tag{7}$$

where the set of inputs X includes input vector $\mathbf{x}_t$; and the set of parameters B includes the parameter vector $\mathbf{b}$. Polynomial regression includes additional input features such as the powers of the original input features and products of two original input features, known as interaction terms.

2.2   Machine Learning Models

The machine learning models place great emphasis on learning directly from the data. In general, no predefined forms of equations or any assumptions are needed for machine learning models. They are more flexible in the sense that they are not constrained to any predefined forms and are free to extract any knowledge and relationship among variables from the data.

To train a machine learning model, an $n \times k_i$ input training matrix, where n is the number of instances and k_i is the number of input features, and an $n \times k_o$ output/response matrix, where k_o is the number of outputs/responses, are usually required. Because of this kind of design, it is very straight forward to train a machine learning model for multivariate, multi-step time series forecasting: simply append more columns to the input matrix and output matrix as needed. The training process usually considers one output feature at a time.

Support Vector Regression. Support Vector Regression (SVR) [6,25], in its simplest linear form, expresses $f(X, B)$ in Eq. 1 as follows,

$$f(X, B) = \langle \mathbf{x}_t, \mathbf{b} \rangle + c, \tag{8}$$

where the $\mathbf{x}_t$, during training, is the instance associated with training output y_t. Please note that $\mathbf{x}_t$ often contains information up until time t, but not at time t. The pair of angle brackets $\langle \rangle$ is the inner product operator. The set of inputs X includes $\mathbf{x}_t$. The set of parameters B includes a parameter vector $\mathbf{b}$ and an intercept c. Optimization is performed to minimize

$$\frac{1}{2}||\mathbf{b}||^2, \tag{9}$$

subject to the constraint of

$$|y_t - (\langle \mathbf{x}_t, \mathbf{b} \rangle + c)| \le \epsilon, \tag{10}$$

that is, the forecasted value must be within a threshold ϵ of the observed value for all training instances. Often times, a non-linear kernel function may be used to transform the training instances into higher dimensional space in order to fit a curve rather than a line. The parameter vector $\mathbf{b}$ may also be expressed as a linear combination of selected training instances, known as support vectors [25].

Neural Networks. Neural Networks (NN) have garnered much attention in recent years, primarily due to the advancement in deep learning research. The standard 3-layer Neural Network, when containing more than one output neuron, may express $g(X, B)$ in Eq. 2 as follows,

$$g(X, B) = f_1(B_1^T f_0(B_0^T \mathbf{x}_t + \mathbf{c}_0) + \mathbf{c}_1), \tag{11}$$

let k_h be the number of hidden nodes, then the set of parameters B includes the $k_i \times k_h$ parameter/weight matrix B_0, the k_h dimensional bias/intercept vector $\mathbf{c}_0$, the $k_h \times k_o$ parameter/weight matrix B_1, and the k_o dimensional bias vector $\mathbf{c}_1$. The two activation functions, f_0 and f_1, output signals from input layer to hidden layer, and from hidden layer to output layer, respectively. Additional hidden layers may be added to a Neural Network and its forecasted/predicted values may be produced in a similar layer-by-layer manner. Since the information can only be passed in a forward manner, and every pair of adjacent layers are completely connected by edges, such Neural Networks are also more precisely called feed forward fully connected Neural Networks.

Long Short-Term Memory Neural Networks. Long Short-Term Memory (LSTM) Neural Network [8] is a type of recurrent Neural Networks designed to work with temporal data. The core of an LSTM NN is an LSTM unit, which may also be viewed as a special layer. The input to an LSTM unit/layer must contain an additional temporal dimension to the standard instances × features input matrix used in other machine learning models. In other words, a training input instance to an LSTM unit/layer contains the temporal evolution of the values of the features.

An LSTM unit/layer contains a cell state that maintains valuable information throughout time. Three gates exist within an LSTM unit/layer that affects the information stored in the cell state: (1) the input gate determines new information that needs to be added to the cell state; (2) the forget gate determines what old information is no longer relevant in the cell state; (3) the output gate determines what output signals to produce based on the contents of the cell state.

Once the final output from an LSTM unit/layer has been obtained at the last time step, then the output may be fed into another Neural Network layer, such as a fully connected layer described in the previous section, and the final forecasted output may be obtained in a similar layer by layer manner described in Eq. 11.

3 Related Work

Univariate time series forecasting of traffic flow is most common in the literature. In [26], a Neural Network was used as a meta-learner trained based on the outputs of ARIMA, MA and exponential smoothing models. The performance was better than ARIMA and a small NN. In [10], the authors' proposed SVR model outperformed SARIMA, exponential smoothing, and a small Neural Network. A work in [15] compared SARIMA, SVR and NN using 15-min resolution data collected over 9 months by 16 sensors from PeMS. The SARIMA model performed the best, but the authors' proposed SVR model ran much faster without losing much accuracies. The NN structure was relatively small in size, and did not perform too well. In [17], both speed and flow data were used to forecast speed, and the performance of LSTM NN was superior to ARIMA, SVR and other Neural Networks. In another work [12], LSTM NN outperformed ARIMA, NN, and Deep Belief Networks. Rainfall data were also included to improve forecasting accuracies.

Multivariate time series forecasting generally relies on using spatially dependent sensor data to improve performance. In [3], a VAR model took advantage of the spatial dependencies between sensors that are on the same freeway and outperformed univariate ARIMA and SARIMA models. In [16], the authors' proposed deep Neural Network built with stacked autoencoders took 5-min resolution traffic flow data during the first two months of 2013 as training data. The deep Neural Network was responsible for learning any temporal and spatial dependencies among the data and its performance was better than SVR and other types of Neural Networks. In another work [29], the authors' proposed LSTM structure took advantage of both spatial and temporal dependencies and outperformed ARIMA, SVR, and other types of Neural Networks. The data were in 5-min resolution, collected during first half year of 2015, from 500 sensors in the 5th Ring (city bypass) in Beijing.

In comparison with our recent previous work in [21], this work improves upon the univariate aspect by including additional univariate forecasting models such as a SARIMA model based on BIC, which generally works better than the SARIMA model in [21] for the southern CA datasets, various regression models, the SVR model, and the increasingly popular LSTM NN. In preliminary testings, we also attempted to improve upon the NN model using *tanh* activation functions found in [21] by using the leaky ReLU [18] activation function which results in higher accuracies. The SARIMA and NN models were the top two performers in [21], and improvements are made upon both in this work. In addition, this work includes a new multivariate experiment to study the effects of spatial dependencies on traffic forecasting. From the statistical models, two seasonal VARMA models and various regression models are included. The machine learning models from the univariate experiment are also used for multivariate traffic flow forecasting and are discussed in more details in the Evaluations section.

4   Evaluations

This section provides details on the datasets, experimental setup, and evaluation results.

4.1   Dataset Description and Preprocessing

The traffic data are obtained from Caltrans Performance Measurement System. This study focuses on southern California (San Diego and surrounding areas), or district 11 as classified by the California Department of Transportation. All data are from major highways, or Mainline (ML) according to PeMS classification. The resolution of the data is 5 min. A total of 373 sensors are chosen in this study. The size of all the data is approximately 1.5 GB. Table 1 contains a summary of the number of sensors selected from each highway. All sensors chosen for this study must contain data for the entire year of 2018.

Table 1. Number of sensors from each highway

Highway	15 N	15 S	5 N	5 S	52 E	52 W	8 E	8 W	78 E	78 W	805 N	805 S
Sensors	38	54	78	79	4	8	21	20	5	11	28	27

The quality control system of PeMS is very robust. If missing data arise due to sensor failure, PeMS automatically imputes the data and provides the imputed data to the users. The users are also given information on the percentage of observed (non-imputed) data across all lanes at any sensor location. In extremely rare occasions, the data provided by PeMS may contain missing data for certain timestamps. In such cases, the missing data are imputed through linear interpolation. All imputed data are included to train models but excluded for performance evaluations.

4.2   Evaluation Metrics

Three normalized evaluation metrics are considered, Mean Absolute Percentage Error (MAPE), Normalized Root Mean Squared Error (NRMSE), and coefficient of determination R^2

$$MAPE = \frac{1}{T}\sum_{t=1}^{T}\left|\frac{y_t - \hat{y}_t}{y_t}\right|, \tag{12}$$

$$NRMSE = \frac{T}{\sum_{t=1}^{T} y_t}\sqrt{\frac{\sum_{t=1}^{T}(y_t - \hat{y}_t)^2}{T}}, \tag{13}$$

$$R^2 = 1 - \frac{\sum_{t=1}^{T}(y_t - \hat{y}_t)^2}{\sum_{t=1}^{T} y_t^2 - \frac{1}{T}(\sum_{t=1}^{T} y_t)^2}, \tag{14}$$

where T is the total number of instances in the evaluation set. Two experiments are conducted, a univariate forecasting experiment and a multivariate one. All testings are performed on the Sapelo2 cluster from Georgia Advanced Computing Resource Center[2].

---

[2] https://gacrc.uga.edu/.

4.3   Problem Analysis and Modeling

Data from the first 8 months of 2018 are used to train models and the last 4 months are used to evaluate the performances. Only data from work days are considered, as weekend data are usually of significant different patterns. Such practice is common is the literature, as can be seen in [15,16]. Furthermore, the evaluation is focused on daytime traffic from 7:00AM to 7:00PM since traffic is most congested and dynamic during daytime. Forecasts are produced for 12 steps ahead, or up to 1 h ahead since the data are in 5-min resolution. A baseline, weekly historical averages computed from the previous 4 weeks, is included to compare against other forecasting models.

Univariate Experiment. In the univariate experiment, a model is trained only with historical data from one particular sensor. The SARIMA $(1,0,1) \times (0,1,1)_{1440}$, denoted as SARIMA in Fig. 1, is commonly used in literature [15,21,23]. The seasonal period is one week ($1440 = 5$ work days per week $\times$ 24 h per day $\times$ 12 five-minute periods per hour). The SARIMA $(5,0,5) \times (2,1,1)_{1440}$, denoted as SARIMA2, is also used and its parameters are found using a grid search like algorithm proposed by [11] based on the BIC criterion [22] on a small subset of the data. Optimization of exponential smoothing parameters is done by minimizing one-step ahead within sample SSE.

The input features of machine learnings and regression models include 12 most recent traffic flow observations, 12 observations from the previous seasonal period (i.e., when forecasting this coming Monday's traffic from 8:00AM to 9:00AM, last Monday's traffic flow data from the same time window are used), 12 historical averages computed from previous 4 weeks (the baseline), and the time of the day. The training output matrix simply includes traffic flow data for the next 12 steps (1 h). All data are normalized between 0 and 1.

Various regression models are considered, including linear regression (Reg), quadratic regression without interaction terms (QuadReg), quadratic regression with interaction terms, also known as response surface regression (RespSurf), and cubic regression without interaction terms (CubicReg). The ν-SVR model, in which the parameter ν controls the number of support vectors, is chosen for its efficiency. The SCALATION implementation is based on the LIBSVM package [4]. The value of ν is set to 0.05 and cost is set to 1.0 through grid search. The NN model consists of a 4-layer structure. The two hidden layers are using leaky ReLU activation functions [18]. The output layer is using the identity activation function. Since 12-step ahead forecasts are desired, 12 NNs are trained, each with a single output neuron representing a particular forecasting horizon. The size of each layer is half the size of its previous layer. By using grid search, the number of training epochs is set to 300, batch size is 32 and the alpha parameter in leaky ReLu is 0.3 (the latter two parameters are also default values in Keras). Optimization is done using the Adam algorithm [13]. LSTM NN uses a very similar set up with NN, except that the first hidden layer is replaced with the LSTM layer and the number of training epochs is only 100. In addition, the inputs to the LSTM NN require an additional temporal dimension, therefore instances in the training input matrix are further grouped weekly for 4 weeks.

Multivariate Experiment. On a given highway, there can be many sensors. It is intuitive that data from traffic sensors in close proximity are spatially dependent. Data from an upstream sensor can provide information on upcoming congestions while data from downstream sensors determines the rate of traffic flow down the road. In this multivariate traffic flow forecasting experiment, traffic sensors on a particular highway are first sorted by either longitude or latitude, depending on the direction of the highway; then they are divided into groups of 3, with a distance of about 5 miles between any neighboring sensors in the group. The central sensor in each group is the focus of forecast, considering data from both the upstream sensor and the downstream sensor. In this experiment, both flow and speed from all 3 sensors are considered and features are generated in a similar manner as in the univariate experiment. Due to the large number of available features, a simple feature selection process is also conducted to extract 72 most useful features, which would be approximately one-third of all the available features. The feature selection is based on a repeated forward selection process that picks the next best feature which optimally improves the overall adjusted R^2 when fitting a linear regression model. The other setups of the multivariate models are kept the same with their univariate counterparts. No multivariate generalization of exponential smoothing is considered in this study; and preliminary results show that response surface regression performs poorly, possibly due to a large number of interaction terms, and is therefore also excluded from the multivariate experiment.

4.4 Forecasting Evaluations

The forecasting evaluation results are aggregated from all 373 sensors using weighted average, since each sensor may have a slightly different number of observed (non-imputed) values in the evaluation set.

In Fig. 1, the performance of univariate forecasting models are compared. Since the baseline is independent of the forecasting horizon (number of steps ahead to forecast), it is denoted as a flat line. In general, the machine learning models produce better forecasting accuracies, with LSTM NN and NN leading in terms of performance. response surface regression also performed well, followed by SVR and other regression models. Exponential smoothing performs well in the initial steps, but its performance quickly degrades and becomes worse than the baseline around step 7 to 9. The SARIMA $(5, 0, 5) \times (2, 1, 1)_{1440}$ generally performs better than the SARIMA $(1, 0, 1) \times (0, 1, 1)_{1440}$ model, perhaps except for forecasting the very first step, depending on the evaluation metrics. It is also worth noting that comparing the time series models with machine learning and regression models may not be completely fair, as the input features differ greatly. Though this point may also be argued as a strength for machine learning and regression models, which have the flexibility of incorporating various input features.

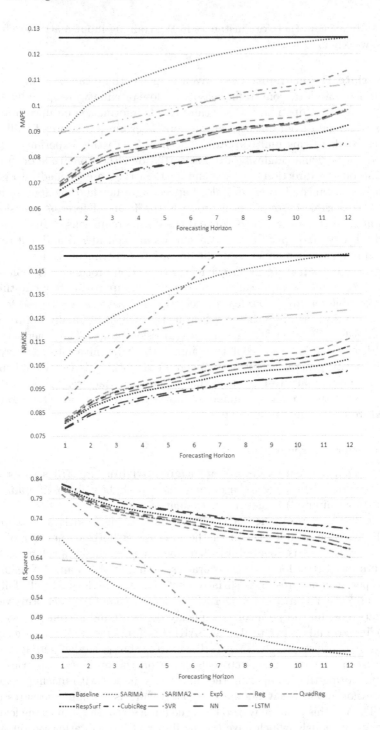

Fig. 1. Univariate models performance comparison

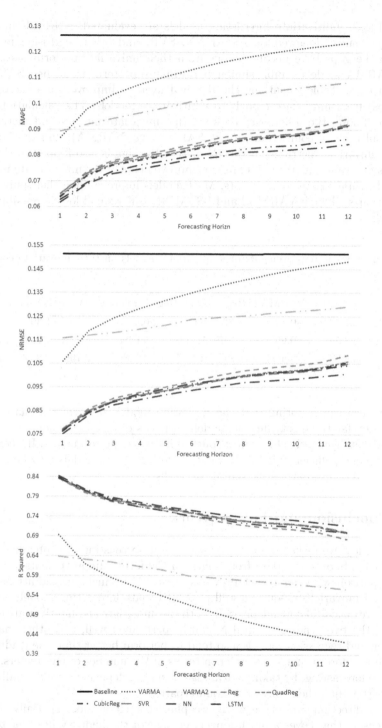

Fig. 2. Multivariate models performance comparison

In Fig. 2, multivariate forecasting models are evaluated. NN leads in overall performance, followed by LSTM NN, SVR, and other regression models, though the gaps have become closer from their univariate performances. The two VARMA models exhibit similar performance patterns from their SARIMA counterparts. Table 2 contains the detailed average improvements (across all 12 steps) from univariate models to multivariate models. The calculations are done by taking the differences between the univariate metrics and metrics of their multivariate counterparts (e.g., SARIMA vs. VARMA), and then divide by the univariate metrics. For the R^2 metric, the sign is also flipped so that all the positive values in Table 2 represent improvements of the multivariate models upon their univariate counterparts. Most models improve upon their univariate counterparts, though VARMA2 and LSTM NN suffer small losses, possibly due to overfitting.

Table 2. Average improvements of multivariate models on their univariate counterparts

	VARMA	VARMA2	Reg	QuadReg	CubicReg	SVR	NN	LSTM
MAPE	2.73%	−0.12%	6.03%	6.06%	6.43%	4.55%	1.30%	−1.07%
NRMSE	2.77%	−0.29%	5.70%	5.51%	5.64%	4.23%	1.75%	−1.41%
R^2	5.34%	−0.46%	4.20%	3.77%	3.79%	2.79%	0.85%	−1.08%

It is also worth noting that the regression models perform reasonably well and are by far the most efficient models in terms of training time. The Neural Networks are generally the slowest, but exhibit great performances. In scenarios where training efficiency is highly valued, the regression models may be considered as viable alternatives to Neural Networks.

5   Conclusion and Future Work

In this study, we focus on multi-step short term forecasting of traffic flow using large amounts of sensor data from southern California. Improvements are made in comparison with our recent previous work [21] by including additional and improved univariate models as well as multivariate forecasting models to take advantage of spatial dependencies. In both the univariate and multivariate experiments, the two types of Neural Networks performed well, and other machine learning and regression models also tend to perform better than the traditional time series models that are simpler in terms of the number of parameters. Multivariate forecasting, by taking advantage of spatial dependencies, generally perform better than their univariate counterparts.

As a direction for future work, we plan to include more spatially dependent sensors that cover a much longer segment of the highway in order to test the performance improvements by relying on spatial dependencies. We are also considering Seq2Seq LSTM NN, which should be more suitable for multi-step forecasting. Furthermore, it is also intuitive to include basic theories that involve speed and distance or even simulation models to help make more accurate forecasts. Many in the data science community in recent years have exclusively relied on deep Neural Networks to learn any knowledge in the data that often feels like a black box. We believe by combining well established theories with the recent advancements in data science, we could more efficiently train better forecasting models.

References

1. Abadi, M., et al.: TensorFlow: large-scale machine learning on heterogeneous systems (2015). https://www.tensorflow.org/
2. Box, G.E., Jenkins, G.M.: Time series analysis forecasting and control. Technical report, DTIC Document (1970)
3. Chandra, S.R., Al-Deek, H.: Predictions of freeway traffic speeds and volumes using vector autoregressive models. J. Intell. Transp. Syst. **13**(2), 53–72 (2009)
4. Chang, C.C., Lin, C.J.: LIBSVM: a library for support vector machines. ACM Trans. Intell. Syst. Technol. **2**, 27:1–27:27 (2011). http://www.csie.ntu.edu.tw/~cjlin/libsvm
5. Chollet, F., et al.: Keras (2015). https://keras.io
6. Drucker, H., Burges, C.J., Kaufman, L., Smola, A.J., Vapnik, V.: Support vector regression machines. In: Advances in Neural Information Processing Systems, pp. 155–161 (1997)
7. Guo, J., Huang, W., Williams, B.M.: Adaptive Kalman filter approach for stochastic short-term traffic flow rate prediction and uncertainty quantification. Transp. Res. Part C Emerg. Technol. **43**, 50–64 (2014)
8. Hochreiter, S., Schmidhuber, J.: Long short-term memory. Neural Comput. **9**(8), 1735–1780 (1997)
9. Holt Charles, C.: Forecasting trends and seasonal by exponentially weighted averages. Int. J. Forecast. **20**(1), 5–10 (1957)
10. Hong, W.C., Dong, Y., Zheng, F., Wei, S.Y.: Hybrid evolutionary algorithms in a SVR traffic flow forecasting model. Appl. Math. Comput. **217**(15), 6733–6747 (2011)
11. Hyndman, R.J., Khandakar, Y., et al.: Automatic time series forecasting: the forecast package for R. No. 6/07, Monash University, Department of Econometrics and Business Statistics (2007)
12. Jia, Y., Wu, J., Xu, M.: Traffic flow prediction with rainfall impact using a deep learning method. J. Adv. Transp. **2017**, 10 (2017)
13. Kingma, D.P., Ba, J.: Adam: a method for stochastic optimization. arXiv preprint arXiv:1412.6980 (2014)
14. Koesdwiady, A., Soua, R., Karray, F.: Improving traffic flow prediction with weather information in connected cars: a deep learning approach. IEEE Trans. Veh. Technol. **65**(12), 9508–9517 (2016)
15. Lippi, M., Bertini, M., Frasconi, P.: Short-term traffic flow forecasting: an experimental comparison of time-series analysis and supervised learning. IEEE Trans. Intell. Transp. Syst. **14**(2), 871–882 (2013)

16. Lv, Y., Duan, Y., Kang, W., Li, Z., Wang, F.Y.: Traffic flow prediction with big data: a deep learning approach. IEEE Trans. Intell. Transp. Syst. **16**(2), 865–873 (2015)
17. Ma, X., Tao, Z., Wang, Y., Yu, H., Wang, Y.: Long short-term memory neural network for traffic speed prediction using remote microwave sensor data. Transp. Res. Part C Emerg. Technol. **54**, 187–197 (2015)
18. Maas, A.L., Hannun, A.Y., Ng, A.Y.: Rectifier nonlinearities improve neural network acoustic models. In: Proceedings of ICML, vol. 30, p. 3 (2013)
19. Miller, J.A.: Introduction to scalation (2018)
20. Okawa, M., Kim, H., Toda, H.: Online traffic flow prediction using convolved bilinear poisson regression. In: 2017 18th IEEE International Conference on Mobile Data Management (MDM), pp. 134–143. IEEE (2017)
21. Peng, H., Bobade, S.U., Cotterell, M.E., Miller, J.A.: Forecasting traffic flow: short term, long term, and when it rains. In: Chin, F.Y.L., Chen, C.L.P., Khan, L., Lee, K., Zhang, L.-J. (eds.) BIGDATA 2018. LNCS, vol. 10968, pp. 57–71. Springer, Cham (2018). https://doi.org/10.1007/978-3-319-94301-5_5
22. Schwarz, G., et al.: Estimating the dimension of a model. Ann. Stat. **6**(2), 461–464 (1978)
23. Shekhar, S., Williams, B.: Adaptive seasonal time series models for forecasting short-term traffic flow. Transp. Res. Rec. J. Transp. Res. Board **2024**, 116–125 (2008)
24. Sims, C.A.: Macroeconomics and reality. Econometrica J. Econometric Soc. **48**, 1–48 (1980)
25. Smola, A.J., Schölkopf, B.: A tutorial on support vector regression. Stat. Comput. **14**(3), 199–222 (2004)
26. Tan, M.C., Wong, S.C., Xu, J.M., Guan, Z.R., Zhang, P.: An aggregation approach to short-term traffic flow prediction. IEEE Trans. Intell. Transp. Syst. **10**(1), 60–69 (2009)
27. Tang, J., Liu, F., Zou, Y., Zhang, W., Wang, Y.: An improved fuzzy neural network for traffic speed prediction considering periodic characteristic. IEEE Trans. Intell. Transp. Syst. **18**(9), 2340–2350 (2017)
28. Winters, P.R.: Forecasting sales by exponentially weighted moving averages. Manage. Sci. **6**(3), 324–342 (1960)
29. Zhao, Z., Chen, W., Wu, X., Chen, P.C., Liu, J.: LSTM network: a deep learning approach for short-term traffic forecast. IET Intell. Transp. Syst. **11**(2), 68–75 (2017)

Towards Detection of Abnormal Vehicle Behavior Using Traffic Cameras

Chen Wang$^{(\boxtimes)}$, Aibek Musaev, Pezhman Sheinidashtegol, and Travis Atkison

The University of Alabama, Tuscaloosa, AL 35487, USA
cwang86@crimson.ua.edu

Abstract. Throughout the world, many surveillance cameras are being installed every month. For example, there are over 18,000 publicly accessible traffic cameras in 200 cities and metropolitan areas in the United States alone. Live video streams provide real-time big data about behavior happening in the present, such as traffic information. However, until now, extracting intelligence from video content has been mostly manual, i.e. through human observation. The development of smart real-time tools that can detect abnormal vehicle behaviors may alert law enforcement and transportation agencies of possible violators and can potentially avoid traffic accidents. In this study, we address this problem by developing an application for detection of abnormal driving behavior using traffic video streams. Evaluation is performed using real videos from traffic cameras to detect stalled vehicles and possible abnormal vehicle behavior.

Keywords: Traffic behavior · Object detection · Object tracking · Anomaly detection

1 Introduction

Each year 20–50 million people are injured or disabled in traffic accidents, and unless action is taken, road traffic injuries are predicted to become the fifth leading cause of death by 2030 [26]. Therefore, a number of intelligent transportation systems have been developed in an attempt to improve the traffic control systems for road safety [1,8,20,33]. One of the promising aspects of such systems is the emergence of traffic video processing systems [30] due to recent advancements in deep learning and computer vision [15,23].

In this paper, a state-of-the-art method for real-time object detection is applied to facilitate the development of an application for determining abnormal vehicle behavior using traffic cameras. Studies show that most traffic accidents are caused by human factors, including drivers' abnormal driving behaviors [27]. Our objective is to determine vehicles whose driving behavior is abnormal compared to the nearby vehicles. Specifically, we want to detect stalled vehicles as well as vehicles that drive either significantly faster or slower than the rest of the traffic.

© Springer Nature Switzerland AG 2019
K. Chen et al. (Eds.): BigData 2019, LNCS 11514, pp. 125–136, 2019.
https://doi.org/10.1007/978-3-030-23551-2_9

The proposed algorithm for detection of abnormal vehicle behavior comprises the following steps:

- Step 1: vehicle detection using live traffic video streams
- Step 2: vehicle tracking using detected vehicles
- Step 3: traffic anomaly detection using tracked vehicles

Vehicle detection is performed using a state-of-the-art method for object detection (You Only Look Once (YOLO)) in each frame of the video stream. The YOLO algorithm generates a list of all detected objects, such as cars and trucks, and their bounding boxes in Step 1. This output is then used as input to the object tracking algorithm based on a Kalman filter in Step 2. The object tracking algorithm associates vehicles with unique IDs, such that each vehicle's path can be tracked through consecutive frames. Finally, the number of frames in which a vehicle appears is used as a relative measure of its speed for anomaly detection in Step 3.

2   Related Work

2.1   Object Detection

Detecting vehicles in a video stream is an object detection problem. Object detection is a core problem in computer vision. Fast, robust object detection systems are fundamental to the success of next-generation video processing systems. Such systems are capable of searching for a specific class of objects, such as faces, people, dogs, airplanes, or vehicles [34].

The types of technologies used for object detection have largely been dictated by computing power. For instance, early systems for object detection in stationary images generally used edge detection and simple heuristics [35]. Modern systems that have access to much more storage and processing power, including GPU accelerators, are able to use large sets of training data to derive complex models of different objects efficiently [24].

Recently, deep convolutional neural networks have made significant advancements in image classification [15] and object detection [9]. Note that object detection is a far more challenging task compared to image classification as it requires more complex approaches to solve. Complexity arises due to the need for the accurate localization of objects. Modern systems approach object localization in either a multi-stage pipeline (RCNN [10], Fast RCNN [9], Faster RCNN [24]) or in a single shot manner (YOLO [23], SSD [19]). The multi-stage pipeline approaches all share one feature in common: one part of their pipeline is dedicated to generating region proposals followed by a high quality classifier to classify those proposals. These methods are very accurate, but come at a high computational cost (low frame-rate); in other words, they are not well suited for real-time vehicle detection.

An alternative way of doing object detection is by combining these two tasks into one network. This is possible because instead of having a network produce region proposals, an image is split into a grid of fixed boxes to look for objects. A single convolutional neural network simultaneously predicts multiple bounding boxes and class probabilities for those boxes. You Only Look Once (YOLO) is a state-of-the-art algorithm among single shot detector approaches for object detection [23]. YOLO trains on full images and is the fastest approach while maintaining high accuracy. This makes it a good fit for real-time vehicle detection.

A growing body of literature are applying YOLO and other related deep learning methods to detecting moving objects using a stationary camera: in [33], YOLO is applied in combination with convolutional neural networks to detect vehicle lane changes; in [7], Histogram of Oriented Gradient (HOG) method is used for pedestrian detection; in [32], a multi-directional CNN method is used for vehicle license plate detection; in [11], a comparison is performed among different deep learning methods for effective detection purposes based on their reliability and repeat-ability. However, very few papers apply the above mentioned methods to detection of abnormal behavior of vehicles on roads.

In the proposed project, YOLO is used in combination with the Kalman filter algorithm to detect the abnormal vehicles, which can provide a new insight for transportation surveillance techniques. In fact, related studies mostly focus on vehicles instead of their trajectories [2,11,25], including vehicle shifting, drifting and platooning rather than detecting speeding cases using a stationary surveillance camera. For the remainder of this paper, a new perspective is presented on separating the camera field of view (FOV) into 2 sections - one for each traffic direction - and analyzing them separately. The model is implemented in MATLAB using the YOLO algorithm for object detection and a Kalman filter for object tracking resulting in a log file with tracked vehicles.

2.2   Object Tracking

Object tracking is one of the fundamental challenges in computer vision [34]. The goal of object tracking is to accurately locate objects of interest based on a series of consecutive video frames. The increasing need for automated video analysis has produced a lot of research on object tracking algorithms for various application areas, including surveillance [3], vehicle tracking [17], robotics [16], and medical imaging [29].

The task of estimating the motion path of an object in successive video frames consists of two steps: object detection and object tracking. As shown in Sect. 2.1, the state-of-the-art approaches in object detection generate bounding boxes for detected objects in addition to the probability of detecting a given class of objects, such as cars, trucks, or persons.

The vehicle tracking algorithms used in intelligent surveillance of traffic data can be classified into four major categories: region-based tracking algorithms, contour-based tracking algorithms, feature-based tracking algorithms and model-based tracking algorithms [13]. The choice of these methods depends on the approach chosen for the object detection and the output generated by it. In this research effort, the output is provided as bounding boxes in consecutive frames, which is why an approximation of a contour-based tracking algorithm is chosen.

2.3 Traffic Anomaly Detection

Previous studies have analyzed the detection of abnormal traffic behavior. [14] utilizes Markov random fields (MRF) and hidden Markov model (HMM) in recognizing abnormal vehicle behavior at intersections. The authors in [14] use a 2-D image to generate an object's moving trajectory and remap it within textures. In [5], the anomalous behavior is detected based on trajectory pattern learning module, such that a cluster of main flow direction (MFD) vectors generates the classification of trajectories. In [28], the author proposes a method based on multi-lane trajectories by comparing dense and free-flowing traffic conditions using Kalman filters. In [18], the temporal outliers on roads are detected based on historical similarity trends.

Most of the outlier detection studies focus on lane changes, hence the speeding case is relatively difficult to capture from trajectory analysis as it does not involve a lot of lane changes. In this paper, the authors introduce a new method of abnormal behavior detection focusing only on the speeding scenarios. In practice, road police has a tendency of pulling over outliers as opposed to over-speeding vehicles in general. Thus, this intuition can be realized by comparing the speed of vehicle under test (VUT) with the nearby vehicles on roads.

3 Proposed Framework

3.1 Vehicle Detection Using Live Traffic Video Streams

The first step of the proposed framework is to detect vehicles that appear in the video. By applying the YOLO algorithm, a detected vehicle - regardless whether it is moving or stationary - will be confined within a bounding box. Next, the video is split into frames at a fixed sampling rate. The vehicle speed can be represented by the number of frames it appears in, such that the more frames it appears in, the slower the vehicle is, and vice versa.

For the surveillance video streams, it is helpful to separate the FOV into three parts as shown in Fig. 1. Segments I and II represent each traffic direction, segment III is the farther end of the camera FOV, where the vehicles become too small to be recognized. This separation is important for the following reasons: (1) to remove the top part of the video where vehicles are too small for detection, so that YOLO's recognition accuracy can be improved, and (2) to reduce the number of objects to process for faster computation.

Note that the location stamp of the camera was hidden from the figure due to privacy reasons.

3.2 Vehicle Tracking Using Detected Vehicles

Using the vehicles detected in the previous step, the goal is to track them through multiple consecutive video frames by associating each vehicle with a unique ID. The tracking of objects is based solely on motion. The motion of each track is estimated by a Kalman filter, which is widely used in object tracking [6,36].

Fig. 1. Camera's field of view is split into regions: segments I and II represent opposite lanes, while segment III is removed as vehicles are too small for object recognition

A Kalman filter is used to predict a vehicle's location in each video frame, and to determine the likelihood of each location being associated with the vehicle's ID. Maintenance of vehicle IDs becomes a critical aspect of the object tracking. A list of vehicle IDs is maintained. In a given video frame, some locations may be assigned to vehicle IDs, while others may remain unassigned. The unassigned vehicle IDs are marked as invisible. An unassigned location is associated with a new vehicle ID.

Each vehicle ID keeps account of the number of consecutive frames where it remained unassigned. If the count exceeds a certain threshold then it is assumed to have left the FOV, so it gets deleted. A MATLAB example script called "Motion-Based Multiple Object Tracking" is used to implement vehicle tracking [22]. The script is modified as follows: the object detection system based on the background subtraction algorithm is replaced with the YOLO detection.

3.3 Traffic Anomaly Detection Using Tracked Vehicles

The proposed method applies the YOLO algorithm combined with a Kalman filter to track the vehicles, then to determine the vehicles whose driving behavior is abnormal compared with their neighbors. The anomaly detection is performed by comparing the speed differences δv between the VUT and its close neighbors. If Δv is significant, then the VUT is considered to be an outlier. An authoritative source, namely the *national speeding index*, is proposed for determining such significance.

Some statistical facts from the index are: (1) in the 2010 census issued by the U.S. Census Bureau [12], the working-age population represented 112.8 million people (36.5% over entire US population); (2) 83% of U.S. adults drive on a daily basis per Gallup statistics [4]; and (3) based on Stanford OpenPolicing [31], police officers pull over more than 50,000 drivers nationwide on a typical day due to speeding. Thus, the speed citation rate c_t can be roughly estimated as:

$$c_t = \frac{50 \cdot 10^3}{83\% \cdot 112.8 \cdot 10^6} \approx 0.05\% \tag{1}$$

Hence, based on the mentioned police report, the threshold can be set at 0.05%.

$$\Delta s \triangleq c_t = 0.05\% \tag{2}$$

4   Evaluation Using Real Data

In this section, two experiments are presented for evaluating the proposed framework. First, stalled vehicles are detected using a 19-min long video from a traffic camera located in Mississippi [21]. In the video a disabled car is located on the shoulder of a road. This is an extreme case of abnormal driving behavior. In the second experiment, vehicles are detected that are either significantly faster or slower than the rest of the traffic using a 53-h long video from a traffic camera also located in Mississippi.

The collected traffic data are processed with a normal distribution function, the resulting distribution curve is shown in Fig. 2.

4.1   Detection of Stalled Vehicles

The previously determined threshold can be applied to the traffic flow such that the average speed citation ratio c_t is fixed at 0.05%. Note that a stalled vehicle becomes a stationary object for a relatively long period of time from the camera perspective, such that it appears in a significant number of frames. If the vehicle is stalled long enough, a large frame number gradually accumulates over time and eventually exceeds the average frame number for all of the passing vehicles. Under these conditions, the pre-defined c_t is no longer applicable, hence a new threshold should be computed for the stalled vehicles. Here in Fig. 2(c), the newly determined threshold is at 1.38%, where the threshold is rolling dynamically indicating different vehicle behavior patterns.

Figure 3 shows a typical abnormal traffic behavior captured by the surveillance camera in Mississippi. The white sedan on the curb is the disabled vehicle, while the white SUV next to it is driving normally; the other two vehicles in the

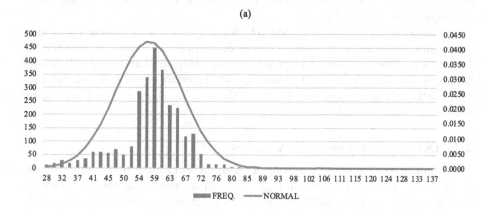

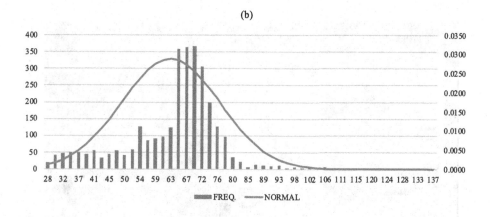

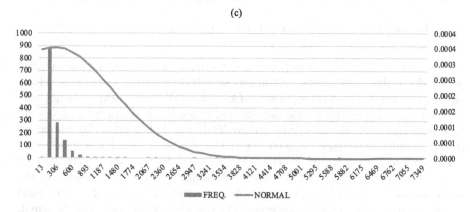

Fig. 2. Normal distribution and frequency histogram of (a) left lane (incoming) traffic flow, (2) right Lane (outgoing) traffic flow, and (c) stalled or disabled vehicle detection.

left lane are also stopped on the road. This situation was maintained a relatively long time before police arrived.

As mentioned earlier, the stalled vehicles in Fig. 3 will be detected over and over again, creating an excessive number of frames and making it possible to detect them. The described detection framework helps find and locate abnormal behaviors in nearly real-time. Combined with the effort of law enforcement and transportation agencies, it may facilitate the development of a safer and more intelligent traffic management system.

4.2 Detection of Abnormal Vehicles

The result of vehicle detection using a snapshot from an actual traffic camera located in Mississippi is shown in Fig. 3. Note that despite the low resolution of the image from the video, all three cars and a truck have been correctly detected, and their bounding boxes have been accurately determined.

Fig. 3. Vehicle detection using a snapshot from an actual traffic camera

For anomaly detection, this paper uses the number of frames for each vehicle in the video: the more number of frames one vehicle is in, the slower this vehicle is, and vice versa. In addition, in some cases the YOLO algorithm was not able to detect vehicles in every successive frame, hence the Kalman filter algorithm helps tracking the vehicles overlooked by YOLO and then adds the number of frames in order to increase the accuracy. In the processed data, a rolling selection compares the VUT with its nearby vehicles on roads, then checks the average number of frames of the total N vehicles including the VUT and calculates the frame-number-ratio j of each vehicle and the average value as in (3). If the ratio is less or larger than the preset threshold Δs, the VUT is considered as abnormal.

$$j(\text{VUT}) = \frac{\text{\# of frames of VUT}}{\text{average \# of frames amongst the neighboring 20 vehicles}} \tag{3}$$

The YOLO and Kalman filter algorithms are integrated in MATLAB resulting in a log file, which contains the raw dataset from the captured vehicles in the surveillance video. The final results are processed using this dataset. Figure 2 shows the normal distribution and frequency histogram results for each traffic direction (referring to Fig. 1 Segments I and II) on roads. From the figure, it is noticeable that most of vehicles are driving at a similar speed while the outliers are the most extreme parts of the curve. In fact, the detected outliers might not be recognized directly by human eyes, thus after validation, the proposed method could be evaluated based on the actual examples of abnormal vehicle behavior.

5   Discussion

The paper presents an initial study of detecting abnormal driving behavior using live video streams. The large variance in road, traffic, weather, sunlight, and other conditions, such as rush hour, heavy rain, etc., may affect the difference in driving speeds of the vehicles on the same road segment. By comparing the speed differences between nearby vehicles under the same conditions during the same time period, we can detect vehicles with abnormal behavior more effectively and convincingly.

The proposed framework successfully detects all stalled vehicles in an actual video from a traffic camera. Note that stalled vehicles represent an extreme case of abnormal behavior, while speeding or slow vehicles are more difficult to detect.

The proposed model uses the number of frames in which vehicles appear for anomaly detection. Specifically, the number of frames in which each vehicle appears is compared with its nearby cars, both before and after it. Thus, the model is sensitive to the accuracy of the number of frames computed for each vehicle's track.

However, even the state-of-the-art method for object detection may generate erroneous vehicle locations. This is possible due to a number of reasons, including the occlusion of a smaller car by a larger vehicle, e.g., a truck. This results in erroneous detection of vehicles by the object tracking algorithm, such that a vehicle's track may suddenly begin in the middle of a road. Similarly, a vehicle's track may end abruptly in the middle of a road due to occlusion.

This problem cannot be resolved by a Kalman filter as it may only detect the missing intermediate locations but not the prior locations. In other words, the vehicle tracking component treats such erroneous tracks as valid. Hence, additional contextual knowledge can be used to filter out such erroneous tracks that either begin or end at an invalid location in the FOV of a traffic camera.

Note that the detection of stalled vehicles is not sensitive to the accuracy of the number of frames as their tracks are significantly longer than the remaining traffic, such that their detection is robust.

6   Conclusion and Future Work

In this paper we propose and evaluate a framework for detecting abnormal vehicle behavior using traffic camera. The proposed approach uses the number of frames, in which a vehicle appears, as a relative measure of the driving behavior. The framework uses a state-of-the-art method for object detection in video frames, then tracks vehicle location through successive frames using a Kalman filter, and finally performs anomaly detection using the number of frames, in which a vehicle appears, as a relative measure of its speed. The proposed approach is applied to detect all stalled vehicles in an actual traffic video stream. In addition, several improvements are suggested to detect other abnormal vehicle behavior, including fast and slow cars.

Based on the presented study, the method of detecting abnormal vehicle behavior using YOLO and Kalman filter is feasible and applicable, while the preset threshold is able to determine whether the vehicle-of-interest is speeding or not. In future work, the authors will be focusing on detecting other abnormal behaviors such as sudden acceleration and deceleration, swerving, and sudden lane changes. In addition, a comprehensive evaluation of the proposed model for such behaviors will be performed. Finally, a comparison between SSD and YOLO will be made to show which single-shot model performs better.

References

1. Abbas, Z., et al.: Short-term traffic prediction using long short-term memory neural networks. In: 7th IEEE International Congress on Big Data, BigData Congress 2018, 2 July 2018 through 7 July 2018, pp. 57–65. IEEE (2018)
2. Asha, C.S., Narasimhadhan, A.V.: Vehicle counting for traffic management system using YOLO and correlation filter. In: 2018 IEEE International Conference on Electronics, Computing and Communication Technologies, vol. 1, pp. 1–6, March 2018
3. Banharnsakun, A., Tanathong, S.: A hierarchical clustering of features approach for vehicle tracking in traffic environments. Int. J. Intell. Comput. Cybern. 9(4), 354–368 (2016)
4. Brenan, M.: 83% of U.S. adults drive frequently: fewer enjoy it a lot, July 2018. https://news.gallup.com/poll/236813/adults-drive-frequently-fewer-enjoy-lot.aspx
5. Cai, Y., et al.: Trajectory-based anomalous behaviour detection for intelligent traffic surveillance. IET Intell. Transp. Syst. 9, 810–816 (2015)
6. Comaniciu, D., Ramesh, V., Meer, P.: Kernel-based object tracking. IEEE Trans. Pattern Anal. Mach. Intell. 25(5), 564–575 (2003)
7. Dalal, N., Triggs, B.: Histograms of oriented gradients for human detection. In: International Conference on Computer Vision & Pattern Recognition, vol. 1, pp. 886–893, June 2005
8. Datondji, S.R.E., et al.: A survey of vision-based traffic monitoring of road intersections. IEEE Trans. Intell. Transp. Syst. 17(10), 2681–2698 (2016)
9. Girshick, R.B.: Fast R-CNN. In: 2015 IEEE International Conference on Computer Vision, ICCV 2015, Santiago, Chile, 7–13 December 2015, pp. 1440–1448 (2015)

10. Girshick, R.B., et al.: Rich feature hierarchies for accurate object detection and semantic segmentation. In: 2014 IEEE Conference on Computer Vision and Pattern Recognition, CVPR 2014, Columbus, OH, USA, 23–28 June 2014, pp. 580–587 (2014)

11. Hosang, J.H., et al.: What makes for effective detection proposals? IEEE Trans. Pattern Anal. Mach. Intell. **38**(4), 814–830 (2016)

12. Howden, L.M., Meyer, J.A.: Contacts between police and the public, 2010. In: 2010 Census Briefs, pp. 1–16, May 2011

13. Weiming, H., et al.: Traffic accident prediction using 3-D model-based vehicle tracking. IEEE Trans. Veh. Technol. **53**(3), 677–694 (2004)

14. Kamijo, S., et al.: Traffic monitoring and accident detection at intersections. IEEE Trans. Intell. Transp. Syst. **1**, 108–118 (2000)

15. Krizhevsky, A., Sutskever, I., Hinton, G.E.: ImageNet classification with deep convolutional neural networks. Commun. ACM **60**(6), 84–90 (2017)

16. Kyriacou, T., Bugmann, G., Lauria, S.: Vision-based urban navigation procedures for verbally instructed robots. Robot. Auton. Syst. **51**(1), 69–80 (2005)

17. Li, C., Hua, T.: Human action recognition based on template matching. Procedia Eng. **15**, 2824–2830 (2011)

18. Li, X., et al.: Temporal outlier detection in vehicle traffic data. In: 2009 IEEE 25th International Conference on Data Engineering, pp. 1319–1322, March 2009

19. Liu, W., et al.: SSD: single shot multibox detector. In: Leibe, B., Matas, J., Sebe, N., Welling, M. (eds.) ECCV 2016. LNCS, vol. 9905, pp. 21–37. Springer, Cham (2016). https://doi.org/10.1007/978-3-319-46448-0_2

20. Lv, Y., et al.: Traffic flow prediction with big data: a deep learning approach. IEEE Trans. Intell. Transp. Syst. **16**(2), 865–873 (2015)

21. MDOTtraffic – Powered by MDOT. https://www.mdottraffic.com/. Accessed 16 Mar 2019

22. Motion-based multiple object tracking - MATLAB & Simulink. https://www.mathworks.com/help/vision/examples/motion-based-multiple-object-tracking.html. Accessed 22 Mar 2019

23. Redmon, J., et al.: You only look once: unified, real-time object detection. In: 2016 IEEE Conference on Computer Vision and Pattern Recognition, CVPR 2016, Las Vegas, NV, USA, 27–30 June 2016, pp. 779–788 (2016)

24. Ren, S., et al.: Faster R-CNN: towards real-time object detection with region proposal networks. IEEE Trans. Pattern Anal. Mach. Intell. **39**(6), 1137–1149 (2017)

25. Ren, S., et al.: Faster R-CNN: towards real-time object detection with region proposal networks. IEEE Trans. Pattern Anal. Mach. Intell. **39**, 1137–1149 (2017)

26. Road Safety Facts - Association for safe international road travel. https://www.asirt.org/safe-travel/road-safety-facts/. Accessed 15 Mar 2019

27. Saiprasert, C., Pattara-Atikom, W.: Smartphone enabled dangerous driving report system. In: 46th Hawaii International Conference on System Sciences, HICSS 2013, Wailea, HI, USA, 7–10 January 2013, pp. 1231–1237 (2013)

28. Sivaraman, S., Morris, B., Trivedi, M.: Learning multi-lane trajectories using vehicle-based vision. In: 2011 IEEE International Conference on Computer Vision Workshops, pp. 2070–2076, November 2011

29. Smal, I., et al.: Multiple object tracking in molecular bioimaging by Rao-Blackwellized marginal particle filtering. Med. Image Anal. **12**(6), 764–777 (2008)

30. Sreekumar, U.K., et al.: Real-time traffic pattern collection and analysis model for intelligent traffic intersection. In: 2018 IEEE International Conference on Edge Computing (EDGE), pp. 140–143. IEEE (2018)

31. The results of our nationwide analysis of traffic stops and searches. https://openpolicing.stanford.edu/findings/
32. Xie, L., et al.: A new CNN-based method for multi-directional car license plate detection. IEEE Trans. Intell. Transp. Syst. **19**, 507–517 (2018)
33. Yao, W., et al.: On-road vehicle trajectory collection and scene-based lane change analysis: part ii. IEEE Trans. Intell. Transp. Syst. **18**(1), 206–220 (2017)
34. Yilmaz, A., Javed, O., Shah, M.: Object tracking: a survey. ACM Comput. Surv. **38**(4), 13 (2006)
35. Yilmaz, A., Li, X., Shah, M.: Contour-based object tracking with occlusion handling in video acquired using mobile cameras. IEEE Trans. Pattern Anal. Mach. Intell. **26**(11), 1531–1536 (2004)
36. Yin, S., et al.: Hierarchical Kalman-particle filter with adaptation to motion changes for object tracking. Comput. Vis. Image Underst. **115**(6), 885–900 (2011)

Author Index

Printed in the United States
By Bookmasters

Printed in the United States
By Bookmasters